Math Mammoth Grade 4
Review Workbook

By Maria Miller

Contents

Introduction

Math Mammoth Grade 4 Review Workbook is intended to give students a thorough review of fourth grade math. The book has both topical as well as mixed (spiral) review worksheets, and includes both topical tests and a comprehensive end-of-year test. The tests can also be used as review worksheets, instead of tests.

You can use this workbook for various purposes: for summer math practice, to keep the child from forgetting math skills during other break times, to prepare students who are going into fifth grade, or to give fourth grade students extra practice during the school year.

The topics reviewed in this workbook are:

- addition, subtraction, graphs, and algebraic thinking
- large numbers and place value
- multi-digit multiplication
- time and measuring
- division
- geometry
- fractions and decimals

In addition to the topical reviews and tests, the workbook also contains many cumulative (spiral) review pages.

The content for these is taken from *Math Mammoth Grade 4 Complete Curriculum*, so naturally this workbook works especially well to prepare students for grade 5 in Math Mammoth. However, the content follows a typical study for grade 4, so this workbook can be used no matter which math curriculum you follow.

Please note this book does not contain lessons or instruction for the topics. It is not intended for initial teaching. It also will not work if the student needs to completely re-study these topics (the student has not learned the topics at all). For that purpose, please consider the *Math Mammoth Grade 4 Complete Curriculum*, which has all the necessary instruction and lessons.

I wish you success with teaching math!
Maria Miller, the author

Addition, Subtraction, and Algebraic Thinking Review

1. Subtract or add in your head.

a. $81 - 72 =$ _____	**b.** $45 + 65 =$ _____	**c.** $160 + 280 =$ _____
$665 - 99 =$ _____	$196 + 99 =$ _____	$54 - 28 =$ _____

2. Write a number sentence using x for the problem. Write the numbers and x in the bar model. Then solve. Do *not* just write the answer.

   Mary had saved $230. Then she bought a flute and some music books. Now she has $38 left. How much did her purchases cost?

   _____

   $x =$ _____

3. Solve $x + 587 = 1,394$.

4. Calculate in the right order.

a. $5 \times (2 + 4)$	**b.** $120 - 20 - 2 \times 0$	**c.** $(80 - 44) + (80 - 34)$
$(50 - 20) \times 2 + 10$	$5 \times 3 + 2 \times 7$	$10 \times (4 + 4) - 4$

5. Which number sentence matches the problem below?

   What is the cost of three $13-hammers when they are discounted by $2 each?

   $3 \times \$13 - \$2$

   $\$13 - 3 \times \$2$

   $(\$13 - \$2) \times 3$

6. How many feet do ten dogs and 20 chickens have in total?
   Write a single number sentence to solve this.

7. Estimate the total cost of the items below using rounded numbers. Don't find the exact cost.

Colored pencils $24.85; number cards $13.95; dice $3.31

8. After spending $15.20 on food and $34.60 on gasoline,
Mom had $70.20 left in her purse.
How much did she have originally?

9. Alberto bought two pairs of skis. One pair cost
$48.90 and the other cost $25 more than the first.
What was the total cost?

Addition, Subtraction, and Algebraic Thinking Test

1. Solve: $2{,}392 + x = 5{,}003$.

2. Calculate in the right order.

a. $(40 + 90) \times 2$	**b.** $(50 - 10) \div (5 - 3)$	**c.** $50 + 10 \times 4 - 20$

3. Find the expression that matches the problem below.

   What is the change you get if you buy seven light
   bulbs for $2 each, and you pay with $20?

   Lastly, solve the problem.

   $7 \times \$2 - \$20$

   $(\$20 - \$2) \times 7$

   $\$20 - 7 \times \$2$

   $7 \times \$20 - \$2$

4. Estimate the total cost using rounded numbers.
   Do *not* find the exact cost.
   A bicycle helmet, $28.95 and TWO flashlights, $14.25 each.

5. Edward had three rolls of plastic. The first one was 10 meters long,
   the second was 2 m shorter, and the third was 5 m longer than
   the first one. What is the total length of the three rolls of plastic?

6. Mark the numbers and the unknown (x or ?) in the bar model. Write an addition or a subtraction with
   an unknown. Solve it.

   A computer program has been discounted by $48,
   and now it costs $67. What was the original price?

   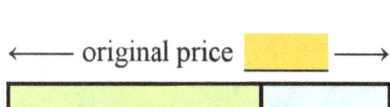
   ←—— original price ——→

   _____

   _____

Large Numbers and Place Value Review

1. Write the numbers.

a. 13 thousand 4 ones
   9 tens
	b. 300 thousand 5 tens
	   6 thousand
		c. 1 million

2. Write the numbers.

    a. 785 thousand 3 hundred   **b.** 70 thousand eight

3. What is the value of the digit 3 in the following numbers?

    a. 21<u>3</u>,047   **b.** 94,0<u>3</u>2

    c. <u>3</u>00,049   **d.** 9<u>3</u>2,255

4. Round these numbers to the nearest thousand and nearest ten thousand.

n	78,974	5,367	2,558	407,409	299,603
rounded to nearest 1,000					
rounded to nearest 10,000					

5. First estimate the result of $5,076 - 2,845 - 675$ by rounding the numbers to the nearest hundred. Then find the exact answer.

Estimation:

Exact answer:

6. Find the missing numbers.

    a. $40,505 = 5 + \underline{\hspace{2cm}} + 40,000$ **b.** $796,000 = 96,000 + \underline{\hspace{2cm}}$

    c. $4,605,506 = 500 + 5,000 + 4,000,000 + 6 + \underline{\hspace{3cm}}$

7. Write < or > between the numbers.

a. 5,406 5,604	**b.** 49530 49553	**c.** 605748 60584

8. Write the numbers in order from the smallest to the greatest.

    5,905,544   95,695   495,644   496,455   145,900   590,554

9. Calculate. Line up the digits in the same place carefully.

    a. $355,399 + 2,455 + 34,200$ **b.** $490,213 - 45,344$

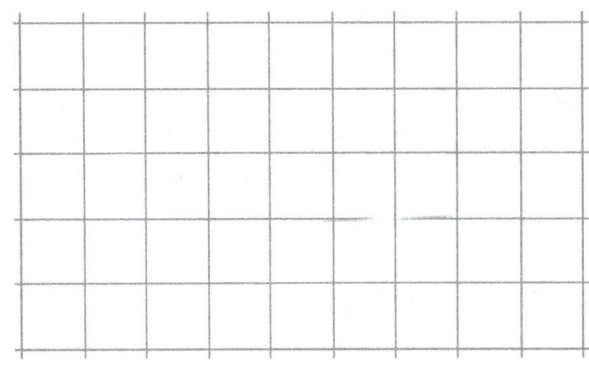 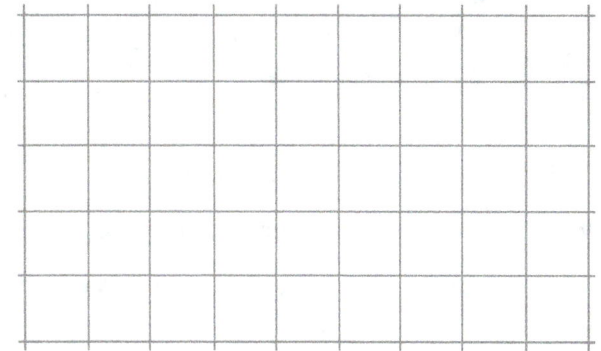

10. A banker puts five hundred $100-bills in a briefcase.
    How much are they worth in dollars?
    Write a multiplication.

11. Mark earns $2,560 in a month.
    How much does he earn in ten months?

    How much does he earn in two months?

    And lastly, how much does he earn in 12 months?

Large Numbers and Place Value Test

1. Write the numbers.

a. 400 thousand 40	b. 4 thousand 5 hundred 60 thousand	c. 200 thousand 7 ones 6 tens

2. What is the value of the digit 8 in the following numbers?

   a. 2**8**0,340   **b.** 294,4**8**7

3. Round the numbers to the underlined place value unit.

a. 516,<u>7</u>64 ≈	**b.** 293,<u>4</u>77 ≈	**c.** 1<u>9</u>6,045 ≈

4. Calculate 225,390 − 17,692.

5. Write the numbers in order from the smallest to the greatest.

                  39,294   3,294   93,294 39,244 399,295

6. The king of Sookiland has 24,000 gold coins in each of
   his three treasuries, plus an additional 1,382 coins in a chest.
   The king of Nootyland owns a total of 78,600 gold coins.
   Which king has more coins? How many more?

7. A charitable organization has one million dollars
   from which they are giving $1,000-grants to
   students. How many students can receive a grant?

Mixed Review 1

1. Write the numbers given in the problem in the bar model. Write x in the bar model for the unknown
   (what the problem asks for). Then write a number sentence using x and solve it.

   Edward bought a raincoat for $23.50 and rubber boots
   for $19.90. He paid and received $6.60 as change.
   What denomination of bill did Edward use to pay?

   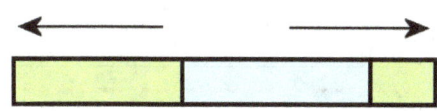

   _____

   $x =$ _____

2. Subtract the money amounts. Check by adding.

   a.

   $$
   \begin{array}{r}
   \$100.00 \\
   - \ \ 43.75 \\
   \hline
   \end{array}
   \qquad + \ \underline{\hspace{3cm}}
   $$

   b.

   $$
   \begin{array}{r}
   \$800.00 \\
   - \ 516.99 \\
   \hline
   \end{array}
   \qquad + \ \underline{\hspace{3cm}}
   $$

3. Which number sentence matches the problem?

   Rakes that cost $15 are discounted by $3.
   You buy four of them. What is the total cost?

   ($15 – $3) × 4

   4 × $15 – $3

   ($15 – $4) × 3

   $15 – 4 × $3

4. Write the numbers in order from the smallest to the greatest: 525,009; 25,925; 5,209; 25,539

5. Either the number you subtract or the number you subtract from is unknown. Solve.

a. $92 - x = 45$	**b.** $x - 566 = 700$	**c.** $900 - x = 267$
$x =$ _____	$x =$ _____	$x =$ _____

6. Solve. Write a number sentence for each problem, not just the answer.

a. Mike had $38. Then Grandma gave him a gift, and
now he has $158. How much did Grandma give him?

_____

b. Jill bought three magazines that cost $4 each with her birthday money. Now
she has $28 left. How much money did she get for her birthday?

_____

c. Greg bought two books that cost $11 each with his birthday money.
He had $60. How much money does he have left?

_____

d. Dad bought each of his three children an ice cream cone that cost $0.60
and an ice cream cone for himself that cost $0.80. How much was the total?

_____

What was his change from $10?

_____

7. Write the numbers in expanded form.

a. 68,056

b. 815,224

8. Round to the nearest dollar.

a. $1.05 ≈ _____   **b.** $7.72 ≈ _____   **c.** $35.17 ≈ _____   **d.** $165.83 ≈ _____

e. $94.90 ≈ _____   **f.** $99.09 ≈ _____   **g.** $99.90 ≈ _____   **h.** $100.56 ≈ _____

9. Estimate the total cost using rounded numbers. Do *not* find the exact cost.

a. A computer math game $19.85; Dictionary $14.90; An encyclopedia on a CD $25.28.

_____

b. 4,000 marker pens at $0.98 each, and 1,000 whiteboard erasers at $1.02 each.

_____

Mixed Review 2

1. Add mentally. You can add in parts (tens and ones separately), or use other "tricks."

a. 56 + 82 = _____	**b.** 29 + 29 = _____	**c.** 69 + 58 = _____
27 + 47 = _____	34 + 58 = _____	25 + 45 = _____
22 + 81 = _____	99 + 45 = _____	72 + 72 = _____

2. Solve in the correct order.

a. $(400 + 200) \times 3 =$ _____ $400 + 200 \times 3 =$ _____	**b.** $10 \times (50 + 10) =$ _____ $10 \times 50 + 10 =$ _____
c. $6 + 9 \div 3 =$ _____ $80 \div 20 \div 4 =$ _____	**d.** $8 \times (300 - 200) - 300 =$ _____ $(70 - 30) \times 4 - 20 =$ _____

3. **a.** Continue this pattern: subtract _____ each time.

700	620	540					

b. Continue this pattern: add 99 each time, starting at 0

0							

4. Write an addition with an unknown (x). Mark the numbers and the unknown in the bar model. Solve.

A shipment of toy cars contained 1,000 cars. Of them, 450 were SUVs, 128 were vans, and the rest were regular cars. How many regular cars were there?

Addition:

Solution: $x =$ _____

5. Add in columns.

a.
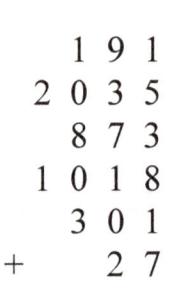
```
        1 9 1
      2 0 3 5
        8 7 3
      1 0 1 8
        3 0 1
  +      2 7
  _____
```

b.
```
          4 3 5
          3 4 9
            2 0
        1 8 1 1
          2 9 4
        9 4 9 3
  +       9 7 6
  _____
```

c.
```
        3 3 7 9 0
        2 3 1 7 6
          7 4 6 3
      6 5 1 0 0 6
            5 1 7
  +          9 9
  _____
```

6. Write the numbers in order from the smallest to the greatest.

   a. 18,399   819,090   8,030   818,939

   b. 52,200   5,220   250,500   520,500

7. Write the numbers.

   a. 284 thousand 1

   b. 50 thousand 50

8. What is the *value* of the underlined digit in the following numbers?

   a. 21<u>2</u>,047

   b. 94,0<u>1</u>2

   c. <u>5</u>00,049

   d. 2<u>4</u>9,255

9. Round the numbers to the nearest hundred.

a. 7,520 ≈ _____	**b.** 2,712 ≈ _____	**c.** 3,953 ≈ _____
d. 354 ≈ _____	**e.** 56,278 ≈ _____	**f.** 293,596 ≈ _____

10. A tablet device was discounted twice: first by $30, then by another $25.
    Now it costs $176. What was the original price?

Multi-Digit Multiplication Review

1. Multiply.

a. $400 \times 3 =$ _____	b. $70 \times 60 =$ _____	c. $90 \times 900 =$ _____
$9 \times 20 =$ _____	$300 \times 11 =$ _____	$100 \times 400 =$ _____

2. Find the missing factors. Think of how many zeros you need.

a. _____ $\times 50 = 4{,}000$	b. $70 \times$ _____ $= 280$	c. _____ $\times 40 = 12{,}000$
_____ $\times 50 = 350$	$7 \times$ _____ $= 2{,}800$	_____ $\times 800 = 64{,}000$

3. Solve the equations.

a. $4 \times 30 = \underline{?} \times 3$	b. $y \times 500 = 250 \times 4$	c. $450 + 350 = \triangle \times 20$
$\underline{?} =$ _____	$y =$ _____	$\triangle =$ _____

4. Solve this problem using **estimation**.
   If you earn $515 weekly, in how many weeks
   will you have earned more than $4,000?

5. Multiply. Estimate the answer on the line.

a. 7×48	b. 6×813	c. 21×18	d. $4 \times 5{,}903$
$\approx$ _____	$\approx$ _____	$\approx$ _____	$\approx$ _____

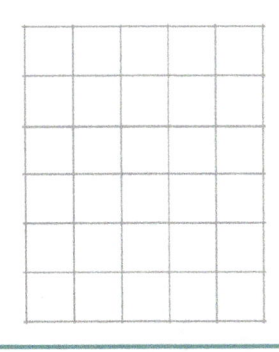  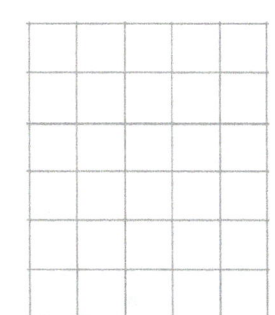 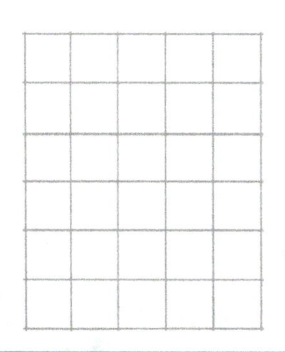

6. Fill in the table.

Roses	1	2	3	4	5	6	7	8
Price	$0.90							

7. Calculate in the right order.

$2 \times 98 - 8 \times 17$

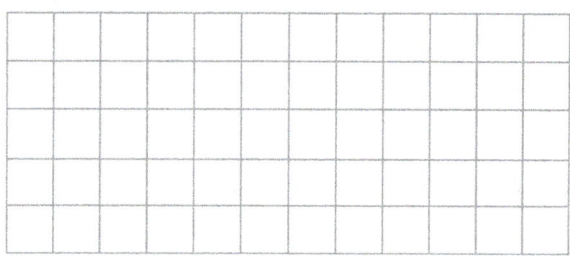

8. Solve.

a. $(1{,}500 - 1{,}000) \times 4 = \underline{\hspace{2cm}}$

b. $(76 + 34) \times 2 \times 0 = \underline{\hspace{2cm}}$

c. $8 \times 2 \times (3 + 2) = \underline{\hspace{2cm}}$

d. $200 \times (500 - 400) = \underline{\hspace{2cm}}$

9. Draw a rectangle with several parts to illustrate the multiplications. You don't have to draw accurately — a sketch is good enough.

a. 8×24

$= \underline{\hspace{1cm}} \times \underline{\hspace{2cm}} + \underline{\hspace{1cm}} \times \underline{\hspace{1cm}}$

$= \underline{\hspace{2cm}}$

b.
$$\begin{array}{r} 3\ 5 \\ \times\ 3\ 9 \\ \hline \end{array}$$

$+$

$\underline{\hspace{3cm}}$

10. Solve. Write a number sentence for each one, not just the answer.

a. A store owner bought 50 boxes of shirts, with 20 shirts
in each box, and each shirt costs $2. What was his total bill?

_____

_____

b. Dad bought 8 boxes of nails for $2.35 a box.
What was his change from $20?

_____

_____

c. Charlene bought five ice cream cones for
$1.50 each. Now she has $12.50 left.
How much did she have originally?

_____

_____

d. A huge roll of wrapping paper costs $45
but it was discounted by $8.
How much do five rolls cost?

_____

_____

11. Solve the problems. You can use the tables to help.

a. A dog can run three
miles in 15 minutes.
How far could it run
in 10 minutes?

b. Seven cans of tuna
weigh 420 g. How
much would ten
cans of tuna weigh?

Multi-Digit Multiplication Test

1. Multiply in your head, part by part.

    a. $4 \times 18 =$ _____   **b.** $7 \times 26 =$ _____   **c.** $3 \times 709 =$ _____

2. One month of tutoring costs $19.85.
   Estimate how much seven months of tutoring would cost.
   (Use a rounded number.)

3. Multiply.

    a. $700 \times 9 =$ _____   **b.** $120 \times 50 =$ _____   **c.** $800 \times 200 =$ _____

4. Find the missing factor.

a. _____ $\times\ 20 = 4{,}000$	**b.** $90 \times$ _____ $= 18{,}000$	**c.** _____ $\times\ 700 = 5{,}600$

5. Solve.

a. $(1{,}500 + 1{,}500) \times 6 =$	**b.** $80 \times (1{,}000 - 400) =$
c. $(34 + 29) \times 0 + 1{,}293 =$	**d.** $(40 - 20) \times 4 + 40 \times 50 =$

6. Multiply.

a.
$$\begin{array}{r} 1\ 5 \\ \times\ 7\ 8 \\ \hline \end{array}$$

b.
$$\begin{array}{r} 7\ 3\ 1 \\ \times\ \ \ \ \ 8 \\ \hline \end{array}$$

c.
$$\begin{array}{r} 5\ 5 \\ \times\ 1\ 9 \\ \hline \end{array}$$

d.
$$\begin{array}{r} 5\ 3\ 0\ 8 \\ \times\ \ \ \ \ \ \ 3 \\ \hline \end{array}$$

7. Calculate $2 \times (48 - 8) \times 17$.

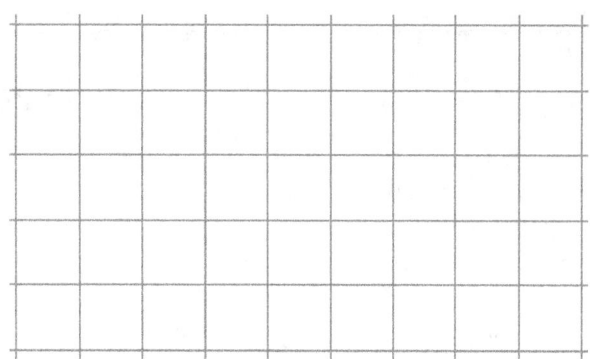

8. Solve.

a. Four super-saver meals cost $12.
How much would seven meals cost?

b. How much will Cindy have left of her $30,
after she buys seven baskets for $2.55 each?

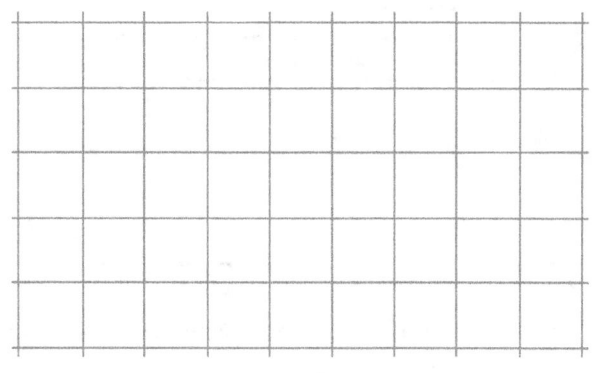

c. Zana went shopping and bought three pairs of
jeans for $12.55 each and a shirt for $8.90.
Now she has $13.45 left. How much money
did she have before she shopped?

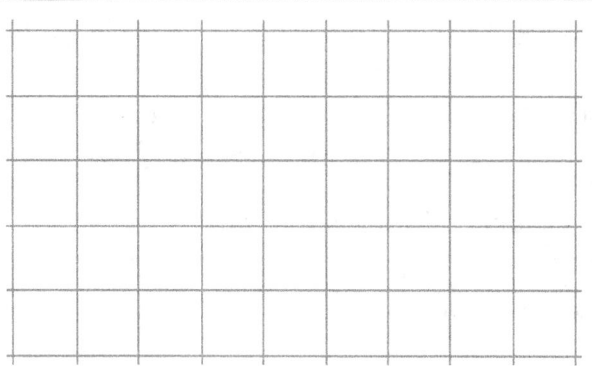

d. Jacob bought five garden hoses with a discount. He paid $100
for them. Without the discount, he would have paid $50 more.
How much did one garden hose cost before the discount?

Mixed Review 3

1. Solve mentally.

   a. 2,000 − (500 + 100) = _____   **b.** 7,000 − (3,200 − 200) = _____

   c. 5,000 + (1,000 − 900 + 100) = _____   **d.** 740 − (550 − 200 + 50) = _____

   e. (900 − 200) − (300 + 200) = _____

   f. 1,000 + (5,000 − 500) + (4,000 − 500) = _____

2. Write the given numbers and *x* in the bar model. Then write another matching subtraction that helps you solve *x*.

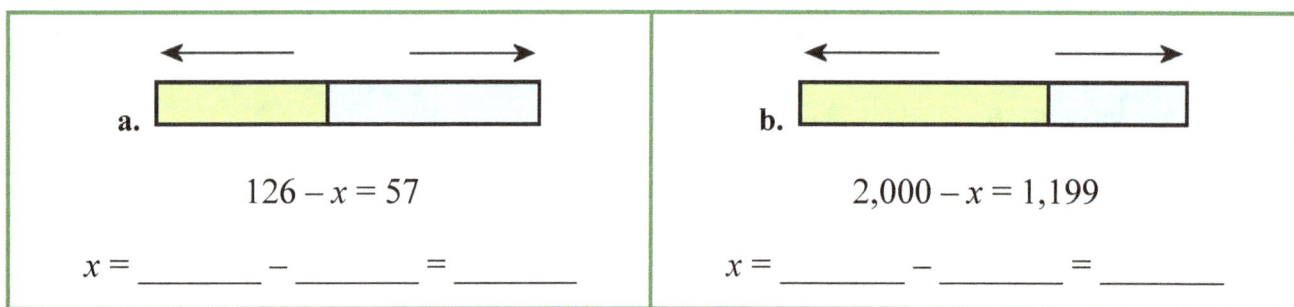

   a.

   126 − *x* = 57

   x = _____ − _____ = _____

   b.

   2,000 − *x* = 1,199

   x = _____ − _____ = _____

3. Round the numbers as the dashed line indicates (to the underlined digit).

a. 44**4**3,920 ≈	**b.** 21**9**,506 ≈	**c.** 617,**0**74 ≈
d. 19**9**,734 ≈	**e.** **3**27,100 ≈	**f.** 81,226 ≈

4. Multiply using the shortcut.

a. 76 × 100 = _____	**b.** 10 × 619 = _____	**c.** 98 × 1,000 = _____
40 × 100 = _____	10 × 2,670 = _____	1,000 × 430 = _____

5. Use the shortcut "backwards" to solve the divisions.

a. 1,560 ÷ 10 = _____	**b.** 700 ÷ 10 = _____	**c.** 21,000 ÷ 1000 = _____
800 ÷ 10 = _____	5,000 ÷ 100 = _____	999,000 ÷ 100 = _____
8,400 ÷ 100 = _____	6,400 ÷ 100 = _____	1,000,000 ÷ 1000 = _____

6. Solve in the correct order.

a. $90 + 15 + 2 \times 7 =$ _____

$90 \times 10 + 120 - 40 =$ _____

b. $500 - 7 \times 70 - 10 =$ _____

$10 \times 7 \times 5 + 100 + 250 =$ _____

7. Compare, and write $<$, $>$, or $=$ in the boxes.

a. 100×26 ☐ 40×70 **b.** $5 + 195$ ☐ 40×5 **c.** 4×72 ☐ 300

8. Mason multiplies wrong. Find what mistake Mason makes each time. Then correct his mistakes.

```
   5
   4 8
 ×   7
 ─────
 2 8 6
```

```
   6
   4 8
 ×   8
 ─────
 3 2 4
```

```
   2
 2 3 9
 ×     3
 ─────
 6 9 7
```

9. Solve.

a. Mick earned $345 by picking strawberries, and Jeanine earned three times as much. How much did they earn in total?

_____

_____

_____

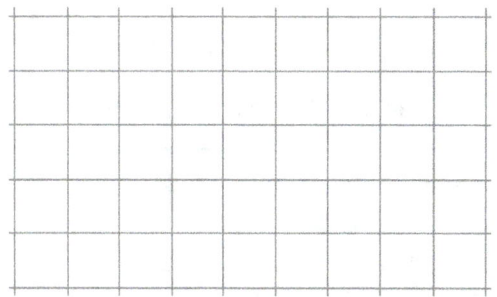

b. A grocery store pays out $145,600 in salaries and $12,390 in other costs *each* month. Calculate its total costs for June, July, and August.

Is the total cost more than half a million dollars?

Mixed Review 4

1. Subtract and compare the results. The problems are "related" – can you see how?

a. 15 − 6 = _____	b. 14 − 8 = _____	c. 12 − 7 = _____
65 − 6 = _____	74 − 8 = _____	82 − 7 = _____
650 − 60 = _____	240 − 80 = _____	1,200 − 700 = _____
250 − 60 = _____	1,400 − 800 = _____	620 − 70 = _____

2. The table lists the sales that Jeanine had for selling homemade dresses.
   Find her total sales over these five weeks.

Week 37	Week 38	Week 39	Week 40	Week 41
$458	$366	$427	$503	$413

3. Fill in.

   a. A car travels at 80 kilometers per hour.

Kilometers								
Hours	1	5	7	9	10	12	15	20

   b. Gary bought fencing for his dog kennel. Four yards cost $36.

Dollars								
Yards	1	2	3	4	5	8	10	15

4. Write the numbers in order.

   a. 5,500   5,005   5,604   5,000   1,554

   _____ < _____ < _____ < _____ < _____

   b. 37,700   73,737   38,707   307,988   3,800

   _____ < _____ < _____ < _____ < _____

5. Add in columns in the grid provided below.

a. $851,091 + 40,510 + 91,576$

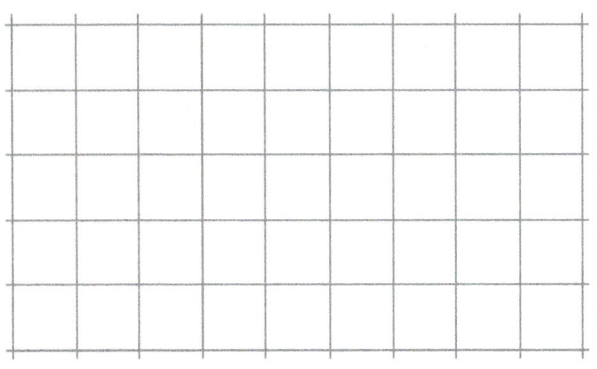

b. $39,312 + 506,636 + 9,382$

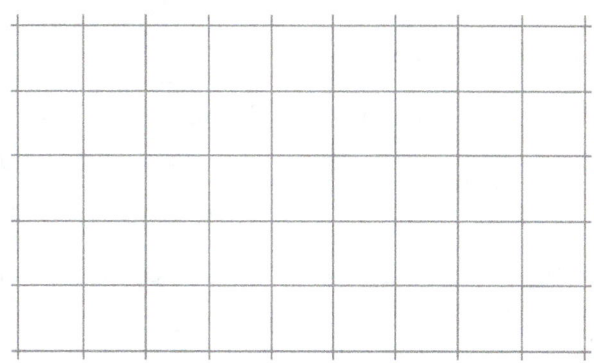

6. Round to the nearest dollar.

a. $\$3.05 \approx$

b. $\$8.32 \approx$

c. $\$25.97 \approx$

7. Beth and Gary helped their mom with a yard sale that they ran for five days. The graph shows how much they earned each day.

a. Estimate their total earnings for these five days.

b. Estimate how much less they earned on their worst day than on their best day.

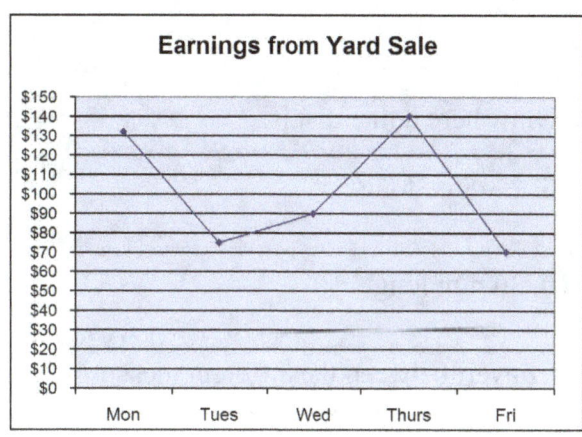

8. Jesse read the whole year's issues of Motorcycle magazine, with 96 pages in each monthly issue. He spent about two-and-a half hours reading each magazine.

a. How many pages did he read?

b. About how long did it take him to read *half* of the magazines?

Time and Measuring Review

1. How much time passes?

   a. From 11:15 p.m. till 6:07 a.m.

   b. From 10:55 till 21:35.

2. A flight that lasts 4 hours 20 minutes took off at 1:50 p.m.
   When does it land?

3. Describe a situation to fit these temperatures.

   a. 25°F

   b. 25°C

4. Draw here a line that is...

   a. 2 3/8 in. long

   b. 36 mm long

5. Convert between the different measuring units.

a. 15 cm = _____ mm	**b.** 4 yd = _____ ft	**c.** 4 m 25 cm = _____ cm
6 cm 8 mm = _____ mm	5 ft 8 in. = _____ in.	8 km = _____ m

6. The sides of a rectangle measure 2 ft 8 in. and 5 ft 9 in.
   What is its perimeter?

7. The distance from Sarah's home to the community center
   is 1 km 400 m. If Sarah goes there *and back* twice a week,
   what distance does she walk in total?

8. Choose the right weight for each thing. Sometimes there are two possibilities.

a. a 5-year old child	**b.** a thick dictionary	**c.** a letter
16 kg 12 lb 34 lb	8 oz 2 kg 12 oz	2 lb 2 oz 20 oz

9. Convert between the units of weight.

a. 7 lb = _____ oz	**b.** 3 T 200 lb = _____ lb	**c.** 2 1/2 kg = _____ g
5 lb 11 oz = _____ oz	7 kg 500 g = _____ g	3,456 g = ____ kg _____ g

10. At his doctor visit, Matthew weighed 23 kg 200 g.
    He had gained 2 kg 350 g since his last visit.
    What did he weigh at his previous visit?

11. Mary gives her cat 6 oz of cat food every day.
    How many days will the 2-lb sack of cat food last?

12. Which is more?

a. 3 gal 11 qt	**b.** 21 fl. oz. 3 cups	**c.** 3 pints 1/2 gallon

13. Convert between units of volume.

a.	**b.**	**c.**
2 L 300 ml = _____ ml	3 qt = _____ pt	4 gal = _____ qt
6,550 ml = ____ L _____ ml	3 qt = _____ cups	2 cups = _____ fl. oz.

14. Alice and Sheila made 3 gallons of punch for a party.
    How many cups of punch does it provide?

A special medicinal honey costs $2.00 per fluid ounce, and Samantha bought a quart. How much did she pay?	**Puzzle Corner**

Time and Measuring Test

1. Supper needs to be ready by 6 p.m. and Mom needs 40 minutes to fix it. She also plans to visit the library for 1 1/2 hours before starting supper. What is the latest time she can go to the library in order to still have enough time to get home to fix supper?

2. Measure the lines below to the nearest eighth of an inch and also in centimeters and millimeters.

   a. _____ in or _____ cm _____ mm

   b. _____ in or _____ cm _____ mm

3. Convert between the different measuring units.

a. 4 lb 2 oz = _____ oz	**b.** 2 L 80 ml = _____ ml	**c.** 7 m 5 cm = _____ cm
76 cm = _____ mm	3 qt = _____ cups	4 kg 500 g = _____ g
5 ft 5 in = _____ in	200 yd = _____ ft	3 T = _____ lb

4. What is the perimeter of a square with sides that are 2 cm 6 mm long each?

5. You can buy liquid dish soap in 1,000-ml bottles or in 400-ml bottles.

   a. How many of the bigger bottles do you need to buy to get 2 liters of dish soap?

   b. How many of the smaller bottles do you need to buy to get 2 liters of dish soap?

6. The dish soap in 1000-ml bottles costs $3.80 a bottle.
   The dish soap in 400-ml bottles costs $2.10 a bottle.

   a. How much does it cost to purchase 2 liters of the first kind of dish soap?

   b. How much does it cost to purchase 2 liters of the second kind of dish soap?

7. Mrs. Higgins put 1 gallon of honey into pint jars. How many jars did she fill?

Mixed Review 5

1. Fill in the missing numbers. Write the area of the *whole* rectangle as a sum of the areas of the *smaller* rectangles. Also find the total area.

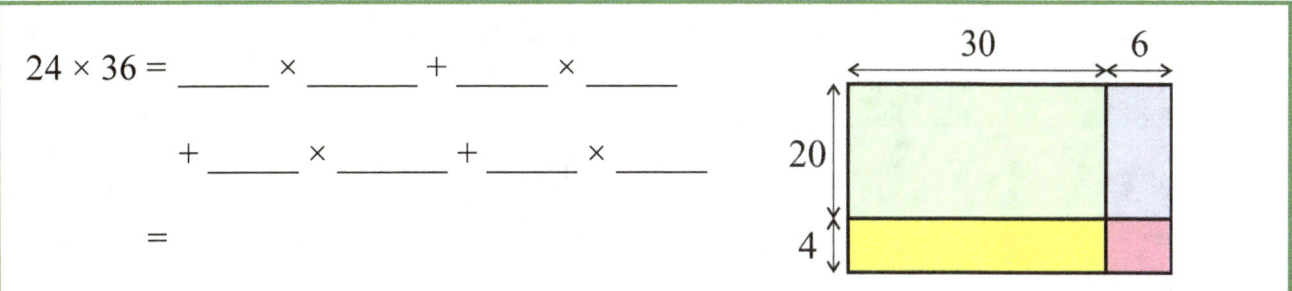

24 × 36 = _____ × _____ + _____ × _____

       + _____ × _____ + _____ × _____

     =

2. Multiply. Estimate the answer on the line.

a. 62 × 29	**b.** 415 × 8	**c.** 57 × 99	**d.** 7 × 667
≈ _____	≈ _____	≈ _____	≈ _____

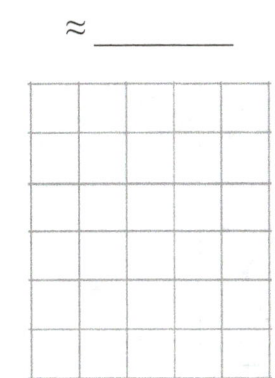 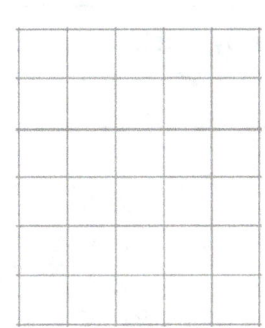 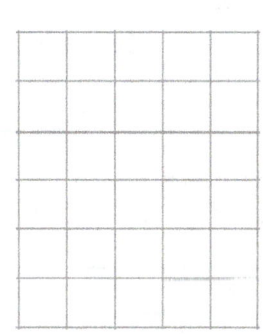 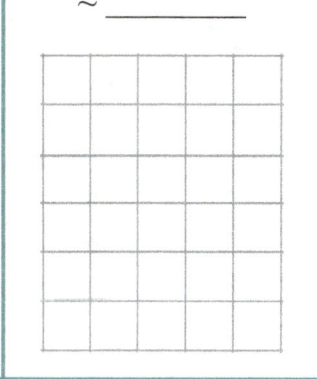

3. Solve.

a. Andrew gets paid $65 weekly for a part-time job. There are 52 weeks in a year but he takes off four weeks each year from work.
How much does he earn in a year?

b. The distance from Steven's home to his job is 24 km. He drives to work and back every day, five days a week.
How many kilometers does he drive in a 5-day work week?

_____

_____

4. Write the numbers *and x* in the bar model. Remember, *x* is the unknown: what the problem asks for.
   Write an addition using the numbers and *x*. Lastly solve.

a. Madison made tortillas. She sold 16 of them to a neighbor, 20 to another, and then had 12 left. How many tortillas did she make?

Addition:

Solution: *x* = _____

b. Emma bought three books for $12 each and a computer part. Her total bill was $73. How much did the computer part cost?

Addition:

Solution: *x* = _____

5. The graph shows the weekly strawberry sales of a small strawberry farm.

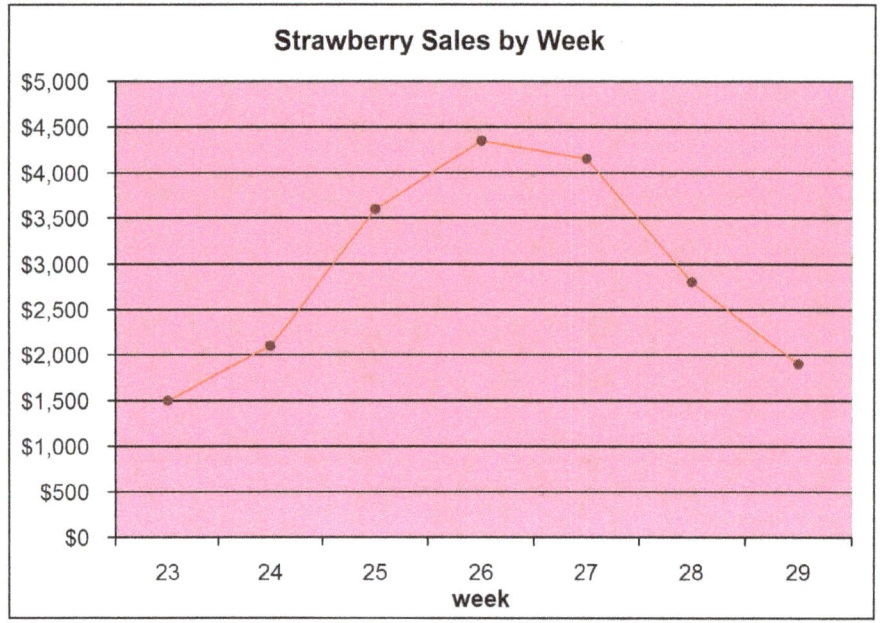

a. What week did the farm sell the most strawberries?

About how much were the sales that week?

b. What week did the farm sell the least strawberries?

About how much were the sales that week?

c. Estimate the total sales for the weeks 25-27.

Mixed Review 6

1. Multiply.

a.
$$\begin{array}{r} 3\ 2\ 5 \\ \times\ \ \ \ \ 3 \\ \hline \end{array}$$

b.
$$\begin{array}{r} 4\ 1\ 9\ 8 \\ \times\ \ \ \ \ \ \ 5 \\ \hline \end{array}$$

c.
$$\begin{array}{r} 8\ 2 \\ \times\ 2\ 4 \\ \hline \end{array}$$

d.
$$\begin{array}{r} 5\ 6 \\ \times\ 7\ 3 \\ \hline \end{array}$$

2. Write an addition sentence using x. Solve.

> The auditorium has 609 seats. Of those, 256 seats are reserved, and the rest are regular. How many seats are just regular?
>
> _____ + _____ = _____
>
> $x =$

3. Do the calculations in the right order.

a. $3 \times (6 + 3) =$ _____	**b.** $(11 - 4) \times 8 + 1 =$ _____
c. $24 - 4 \div 2 =$ _____	**d.** $60 - 1 \times 7 + 12 \div 3 =$ _____
e. $(22 - 16) \times 3 + 3 =$ _____	**f.** $72 \div (6 + 6) - 5 =$ _____

4. Round these numbers to the nearest hundred.

a. $555 \approx$ _____	**b.** $8,889 \approx$ _____	**c.** $351,931 \approx$ _____
d. $64 \approx$ _____	**e.** $244,295 \approx$ _____	**f.** $38,009 \approx$ _____

5. Write the numbers.

   a. three hundred five thousand two hundred

   b. forty thousand thirty-three

6. Place commas into the numbers. Compare. Write either < or > in between the numbers.

a. 7 23050 699099	**b.** 3 2 2 3 2 0 3 2 2 3 2 2
c. 6 9 2 1 5 9 6 9 2 1 9 6	**d.** 1 4 0 0 0 0 1 4 1 0 0
e. 1 1 3 9 9 9 1 1 5 3 9 9	**f.** 8 3 6 4 9 6 8 8 4 8 2

7. Multiply.

a. 100 × 11 = _____	**b.** 18 × 10 = _____	**c.** 100 × 920 = _____
19 × 10 = _____	4,000 × 200 = _____	32 × 2,000 = _____
3,000 × 40 = _____	88 × 100 = _____	400 × 22 = _____

8. Multiply the money amounts in parts. Give your answers in dollars.

a. 6 × 30¢ = _____ ¢ = $_____	**b.** 5 × 84¢ = _____¢ + _____¢ = $_____
c. 6 × $1.70 = _____ + _____ = $_____	**d.** 3 × $2.80 = _____ + _____ = $_____

9. Solve the word problems.

a. Mary planted 22 rows of corn with 38 plants in each row. *Approximately* how many corn plants are there?

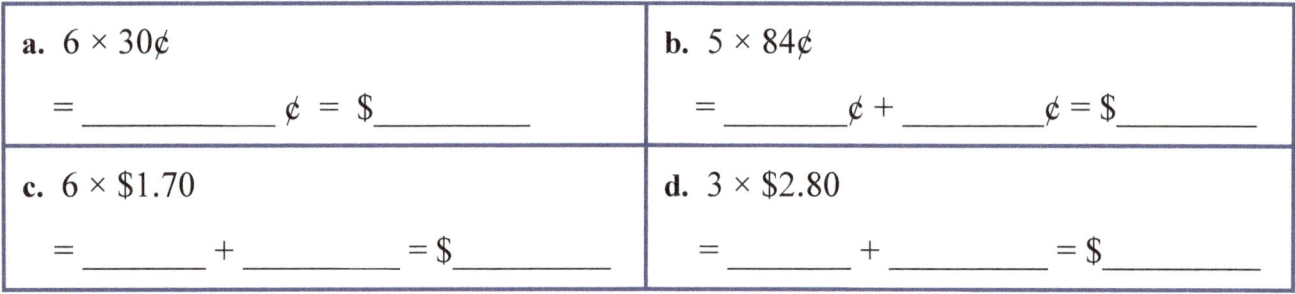

b. Jack had his car fixed. He had to pay for seven hours of work, at $8.20 per hour. He paid with a $100 bill. What was his change?

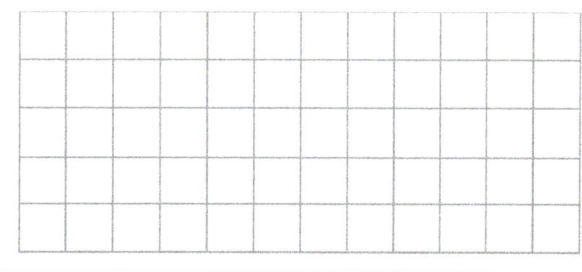

_____

_____

Division Review

1. Solve.

a.	b.	c.
$20 \div 10 + 15 =$ _____	$(200 + 100) \div 5 =$ _____	$10 \times 12 + 40 \div 10 =$ _____
$20 \times 10 + 15 =$ _____	$200 + 100 \div 5 =$ _____	$10 \times (12 + 40) \div 10 =$ _____

2. Solve mentally.

a. $3,100 \div 100 =$ _____	**b.** $240 \div 20 =$ _____	**c.** $4,200 \div 600 =$ _____
$450 \div 10 =$ _____	$800 \div 40 =$ _____	$3,200 \div 80 =$ _____

3. Solve.

a.	b.	c.
$45 \div 6 =$ _____ R _____	$12 \div 7 =$ _____ R _____	$31 \div 4 =$ _____ R _____
$46 \div 6 =$ _____ R _____	$27 \div 8 =$ _____ R _____	$56 \div 9 =$ _____ R _____

4. Divide and check your work.

a. $708 \div 3$ Check:

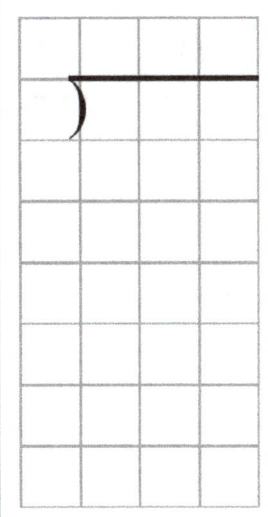

b. $1,504 \div 8$ Check:

5. Divide and check your work.

a. 392 ÷ 5   Check:

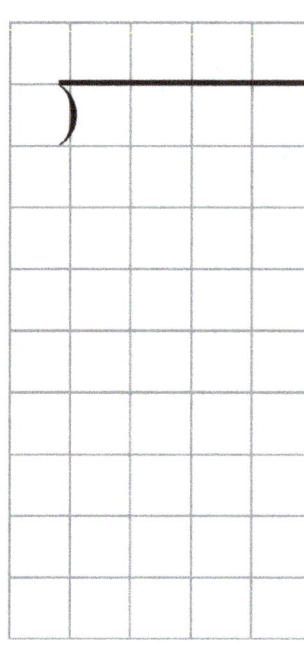

b. 2,845 ÷ 6   Check:

6. Harry has 288 seashells, and Timmy has one fourth
   of that amount. How many does Timmy have?

7. Mark's 70 candles were packaged into boxes.
   Twelve candles fit in each box.

   a. How many boxes became full?

   b. How many candles were in the box that was not full?

8. Four yards of material cost $38.88.
   How much does one yard cost?

9. John's test scores were 92, 85, 89, 75,
   and 89. Find his average score.

10. Mark an X if the number is divisible by 3, by 5, or by 10.

Number	13	40	57	135	354	2,380
Divisible by 3						
Divisible by 5						
Divisible by 10						

11. Answer with "yes" or "no," and give a reason.

a. Is 7 a factor of 64? _____ , because _____ .	**b.** Is 98 a multiple of 2? _____ , because _____ .
c. Is 76 divisible by 8? _____ , because _____ .	**d.** Is 30 a factor of 30? _____ , because _____ .

12. Check if these numbers are primes or composites. Use the divisibility rules for 2, 3, and 5 to help you.

a. 87 is prime/composite If composite: 87 = ____ × ____	**b.** 89 is prime/composite If composite: 89 = ____ × ____	**c.** 91 is prime/composite If composite: 91 = ____ × ____

13. Find all the factors of the given numbers.

a. 24 Check 1 2 3 4 5 6 7 8 9 10 factors: _____	**b.** 27 Check 1 2 3 4 5 6 7 8 9 10 factors: _____
c. 66 Check 1 2 3 4 5 6 7 8 9 10 factors: _____	**d.** 75 Check 1 2 3 4 5 6 7 8 9 10 factors: _____

Puzzle Corner   Imagine you divided all the numbers from 1 to 100 by 6. Which those numbers would have a remainder of 5, when divided by 6?

Division Test

1. Solve.

a.	b.	c.
13 ÷ 4 = _____ R ____	33 ÷ 7 = _____ R ____	40 ÷ 12 = _____ R ____
13 ÷ 5 = _____ R ____	45 ÷ 8 = _____ R ____	67 ÷ 9 = _____ R ___

2. If seven meters of material costs $42,
   what would five meters of material cost?

3. Jesse had saved $350. He spent $\frac{3}{5}$ of that to buy a camera.

   How much did the camera cost?

4. José sold 2/3 of his 1,200 bricks to his neighbor.
   Then, José sold another 150 bricks.
   How many bricks does he have left now?

5. Solve. Check by multiplying.

a. 565 ÷ 5   Check:

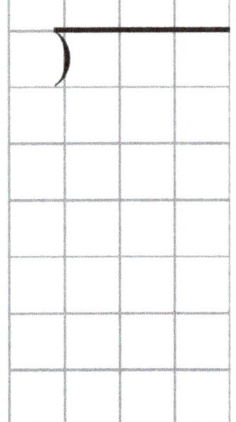

b. 3,664 ÷ 8   Check:

6. Find all the factors of the given numbers.

a. 28	b. 13
factors:	factors:
c. 32	d. 76
factors:	factors:

7. Joe shared 125 pencils between seven children, as equally as he could.

How many did each get?

How many were left over?

8. Ashley found four different pairs of dress shoes in a store, with prices of $39, $45, $63, and $41. What is their average price?

9. Is 924 divisible by 7? Explain why or why not.

10. Place the numbers 10, 20, and 30 *and* parentheses into the expression below so that the answer is more than three hundred. Don't forget the parentheses!

_____ − _____ × _____ = _____

Mixed Review 7

1. Subtract in columns. Check by adding.

a.	Add to check:	b.	Add to check:
7 0 1 6 − 2 7 3 2	+ _____	5 1 8 0 0 − 2 7 3 2	+ _____

2. Calculate in the right order.

a. $7 \times (9 + 3) =$ _____	**b.** $5 + 5 \times 6 \div 2 =$ _____	**c.** $90 + 20 \times 50 =$ _____
$91 - 1 - 20 \div 2 =$ _____	$90 - 3 \times (7 + 5) =$ _____	$(600 - 150) \div 5 =$ _____

3. First estimate the result by rounding the numbers to the nearest hundred. Then find the exact answer.

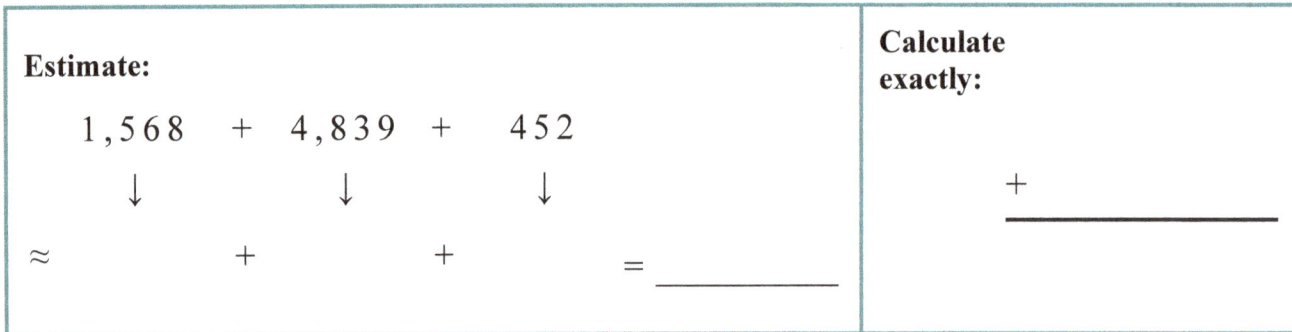

Estimate:

1,568   +   4,839   +   452

↓   ↓   ↓

≈   +   +   = _____

Calculate exactly:

+

_____

4. Subtract mentally from whole thousands.

a. $4,000 - 2 =$ _____	**b.** $7,000 - 10 =$ _____	**c.** $2,000 - 100 =$ _____
$4,000 - 40 =$ _____	$10,000 - 30 =$ _____	$7,000 - 300 =$ _____
$4,000 - 9 =$ _____	$1,000 - 9 =$ _____	$10,000 - 600 =$ _____

5. Write the numbers. The parts are in scrambled order.

a. 6 tens 34 thousand 8 ones 2 hundred	**b.** 800 thousand 4 tens 6 ones	**c.** 7 hundred 6 thousand 400 thousand 8 tens

6. One foot is 12 inches. Convert between feet and inches.

a. 3 ft = _____ in	**b.** 2 ft 5 in = _____ in	**c.** 9 ft 2 in = _____ in
9 ft = _____ in	7 ft 8 in = _____ in	10 ft 11 in = _____ in

7. The expressions are supposed to be equal, but something is missing. Fill in the missing numbers.

a. $5,400 = 90 \times$ _____

b. _____ $\times 20 = 8 \times 40$

c. $7 \times 49 +$ _____ $= 8 \times 49$

d. $24,000 = 300 \times$ _____

e. $7 \times 13 = 5 \times 13 +$ _____

f. _____ $- 500 = 5 \times 200$

8. Solve.

a. Roger has been saving $45 each week to buy himself a laptop for $399. How many weeks will it take? (*Use estimation!*)

Now calculate exactly. How much will he have left over after buying it?

b. Katie teaches a crafts class that has 23 children. For the next meeting she needs to get 10 cm of string, three sheets of paper, and two rolls of tissue paper for each child.
Write down her list of needed supplies.

c. Mark gave half of his 100 marbles to James, and James gave half of those to Greg. How many marbles did James have left after that?

Mixed Review 8

1. **a.** The beach is 1,200 meters long and 1/6 of that is for boat access. How many meters of the beach are not accessible by boat?

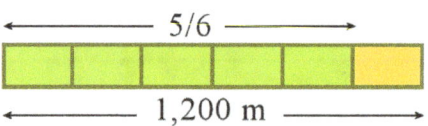

b. Two-thirds of a group of students are girls. There are 11 boys. How many girls are there?

What is the total number of boys and girls?

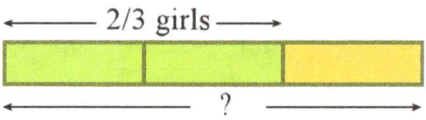

2. A baker charted how much he spent on flour from January through May.

Use rounded numbers, and estimate:

a. *About* how much more did he spend in May than March?

b. *About* how much did he spend in total for March, April, and May?

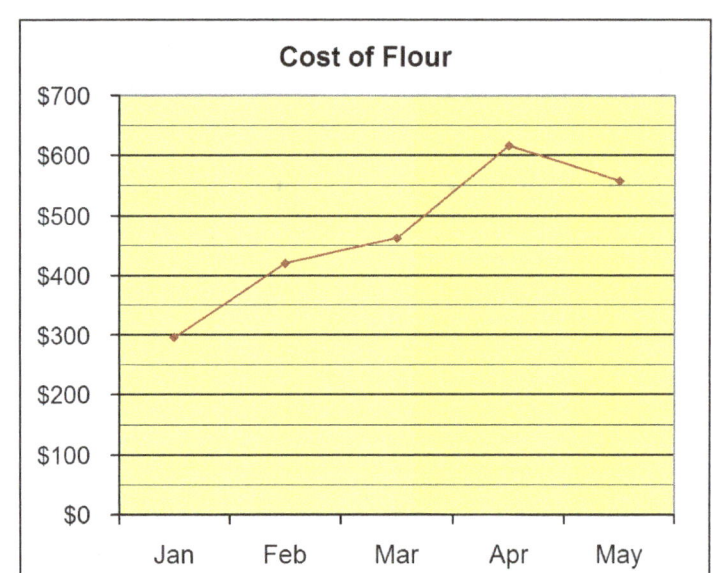

3. Solve.

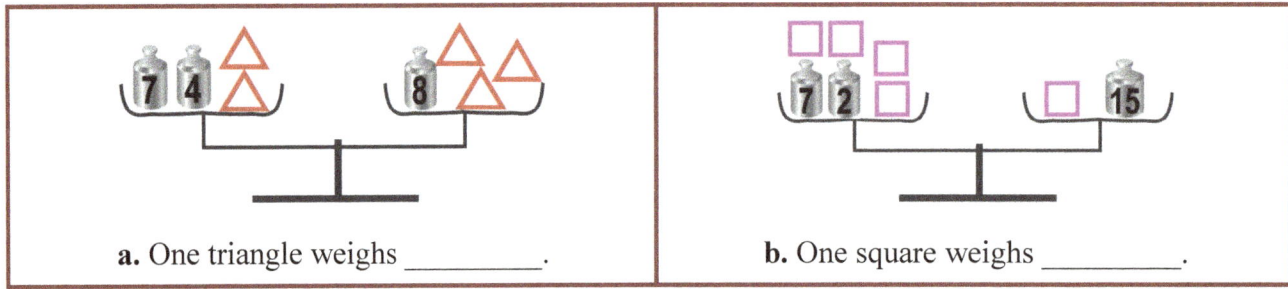

    a. One triangle weighs _____.   **b.** One square weighs _____.

4. Convert.

a. 1 kg 300 g = _____ g	**b.** 3 lb = _____ oz	**c.** 7,500 g = _____ kg _____ g
4 kg 20 g = _____ g	7 lb = _____ oz	4 lb 8 oz = _____ oz

5. Fill in the tables.

Minutes	1	5	6	7	10
Seconds					

Days	1	3	6	10
Hours				

6. Explain in which order to do the operations in the problem on the right. $(5 + 39) \div 4 - 2 \times 2$

First do _____, which equals _____. Then, _____ that answer by _____.

This leaves _____. Then, do _____ = _____.

Lastly, _____ that from _____. The answer is _____.

7. How much time passes? You can use subtraction.

a. From 2:42 p.m. till 7:36 p.m.	**b.** From 3:39 p.m. till 11:03 p.m.	**c.** From 8:45 till 17:09.
h m − h m ————————	h m − h m ————————	h m − h m ————————

8. Convert between the measures of volume.

a.	b.	c.	d.
2 pt = _____ C	1 qt = _____ C	6 L = _____ ml	2 L 560 ml = _____ ml
2 C = _____ oz	2 gal = _____ qt	1/4 L = _____ ml	1,300 ml = _____ L _____ ml

9. Answer the questions.

a. What months could you go sledding?

b. Find the three coldest months of the year.

c. What is the difference between April and July minimum temperatures?

Month	Minimum temp. (°F)
Jan	-35
Feb	25
Mar	0
Apr	28
May	40
Jun	55

Month	Minimum temp. (°F)
Jul	65
Aug	60
Sep	25
Oct	-4
Nov	-20
Dec	-30

Geometry Review

1. A farmer has 45 feet of fencing to fence in a chicken yard. He is only using the fencing on three sides of the yard, and the fourth side will be the wall of the chicken house (10 ft).

   How long are the sides of the yard?

   Side 1:

   Side 2:

   Side 3:

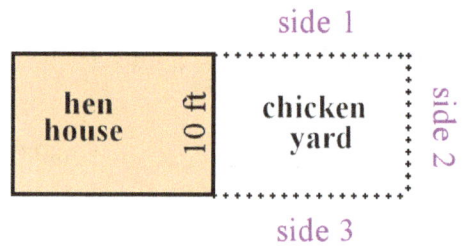

2. Find the area of the shape.

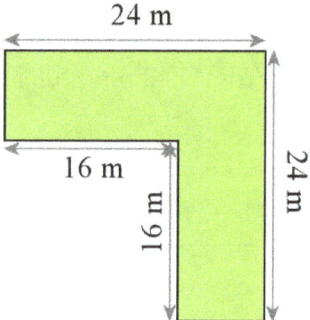

3. Measure the angles.

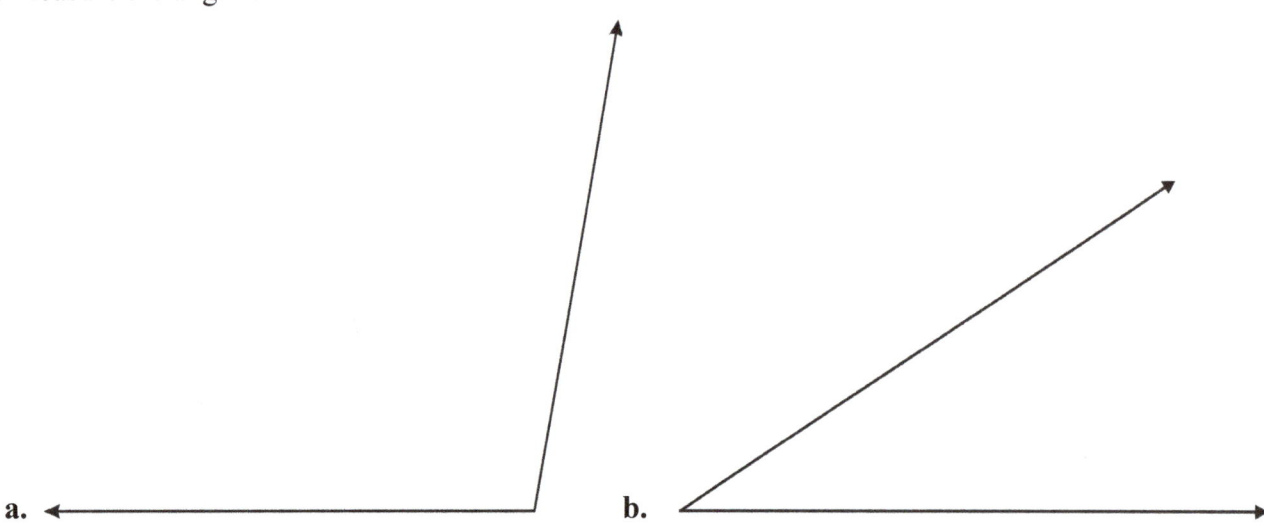

a.

b.

4. Label each angle as acute, obtuse, or right.

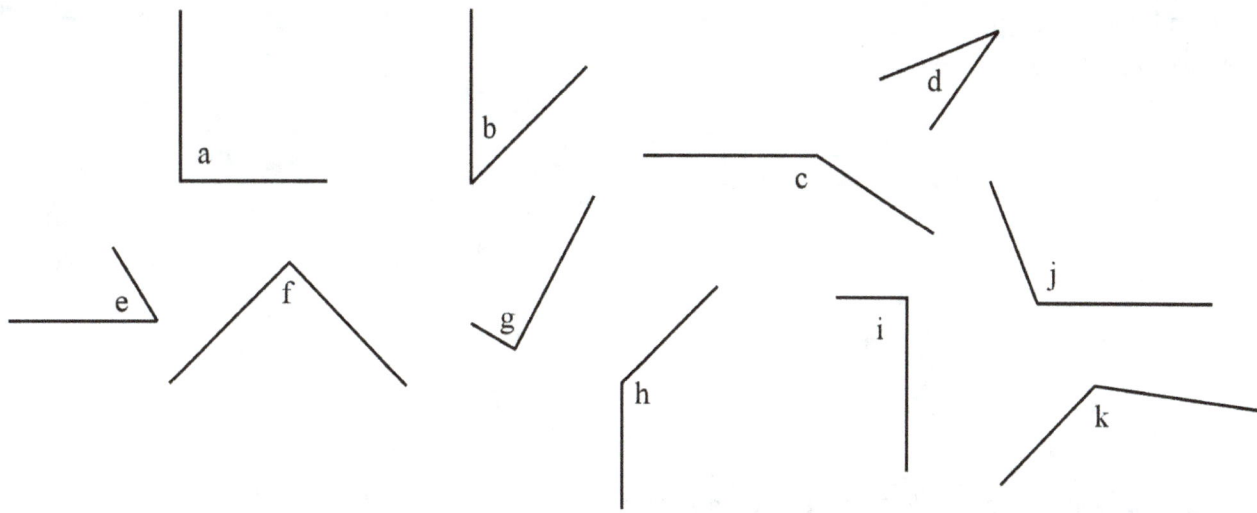

5. Draw an angle that measures 65°.

6. Figure out the unknown angle measure.

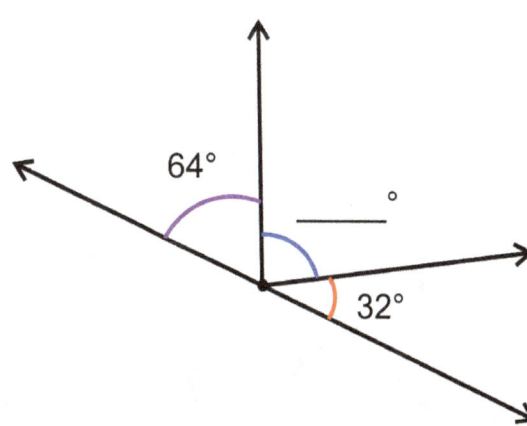

64°

°

32°

7. Draw a line that is perpendicular to this line and goes through the given point.

8. Find the task that is possible to do, and complete it.

   a. Draw a rectangle with one 110° angle.

   b. Draw a triangle with one 110° angle.

   c. Draw an acute angle that measures 110°.

9. **a.** Draw any obtuse triangle.

   b. Measure all the angles of your triangle.

10. Draw ONE **diagonal** (a line from corner to corner) into this square.
    You will get two triangles.
    What kind of triangles are they?

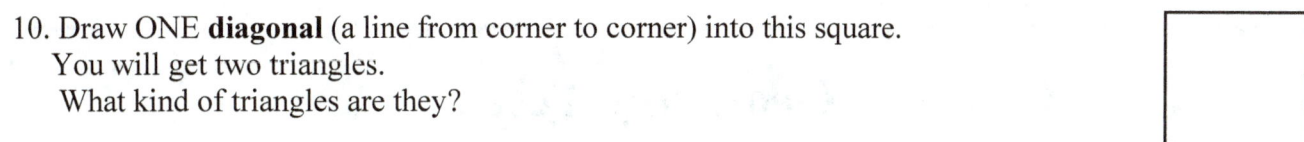

11. Find rays, lines, and line segments that are
    either parallel or perpendicular to each
    other. Use ∥ for parallel and ⊥ for perpendicular.

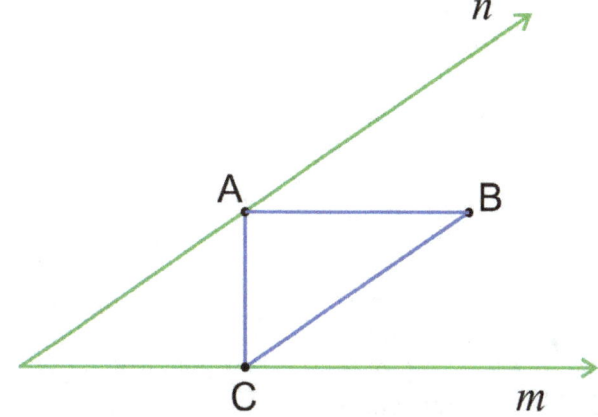

12. Draw any parallelogram.

*Hint: First, draw a line. Then
draw a line that is parallel to
that line. Then draw a third line
that intersects the other two.*

13. Are these figures symmetrical? Draw a symmetry line or lines in the ones that are.

  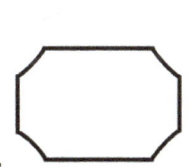  

a.   b.   c.   d.   e.

Geometry Test

1. Draw a 75° angle.

2. Measure this angle.

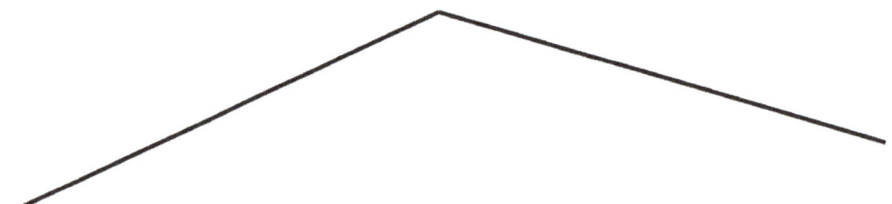

3. What is the measure of the angle *x*?

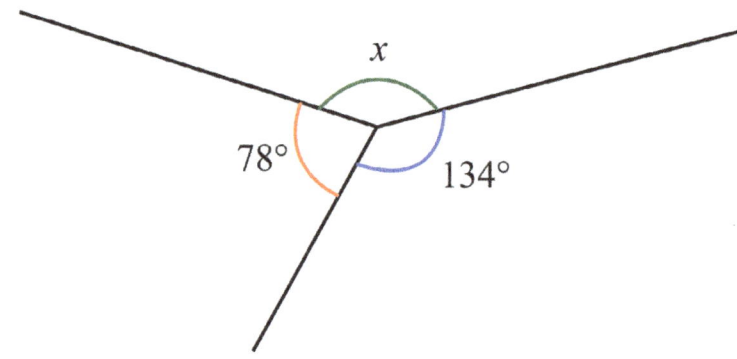

4. **a.** What is this shape called?

   b. Draw a diagonal into it (a line from one
   corner to another) so that you
   will get *two obtuse* triangles.

   c. Measure the perimeter of one of the obtuse
   triangles in centimeters and millimeters.

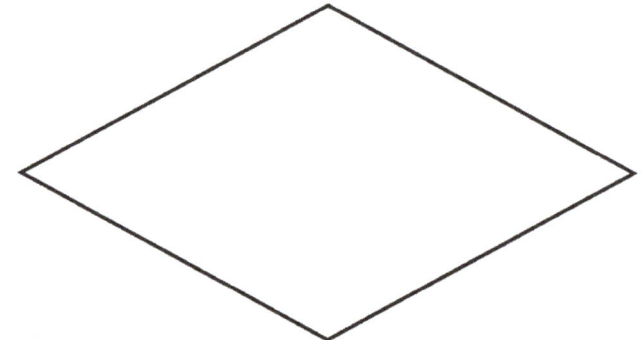

5. Sketch a quadrilateral that has ONE
   right angle, and the other angles
   are not right angles.

6. **a.** Draw a right triangle.

   b. Measure all the angles of your triangle.

7. Classify these triangles according to their angles.   **a.**    **b.**

8. Find the area of the *shaded* area
   in square feet.

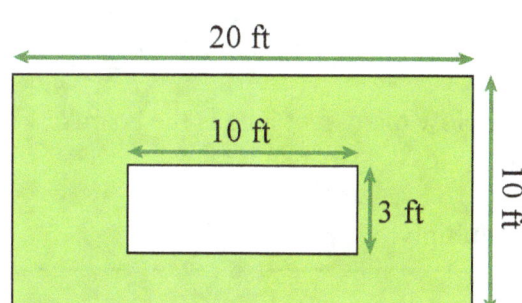

Mixed Review 9

1. Convert between the measuring units.

a. 3 kg = _____ g	**b.** 5 L = _____ ml	**c.** 9 km = _____ m
7 kg 400 g = _____ g	2 L 60 ml = _____ ml	4 km 250 m = _____ m

2. Ed drank 1/2 liter out of a full 2-liter bottle of water.

   How much is left now?

   Then the rest of it was poured into 250-ml glasses.
   How many glasses got filled?

3. Convert between the measuring units.

a. 4 ft = _____ in	**b.** 3 C = _____ fl. oz.	**c.** 4 lb = _____ oz
6 ft 2 in = _____ in	2 gal = _____ qt	7 lb 9 oz = _____ oz

4. Which is more water to drink: a 1-pint bottle, or a glassful of 12 ounces?
   How much more is it?

5. A board is 96 inches long.

   a. How long is 1/12 of the board?

   b. How about 3/12 of the board?

   c. How many feet long is the board?

6. Multiply.

a.	b.	c.	d.
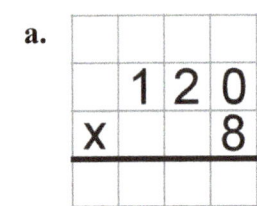	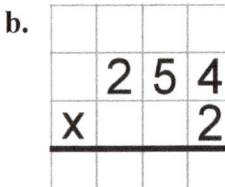	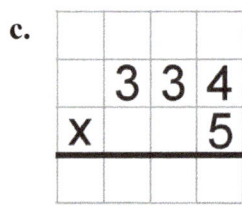	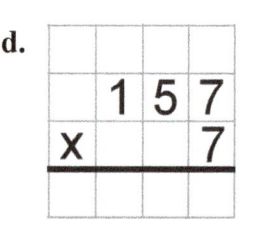

7. Maria studied the weight of 1-year old roosters. She went to a farm and weighed 20 roosters. The numbers below are their weights, in *ounces*.

96, 94, 90, 101, 84, 102, 101, 95, 108, 113, 87, 95, 97, 84, 90, 99, 89, 93, 92, 100.

a. Make a bar graph. Draw the bars touching each other (no gaps between the bars).

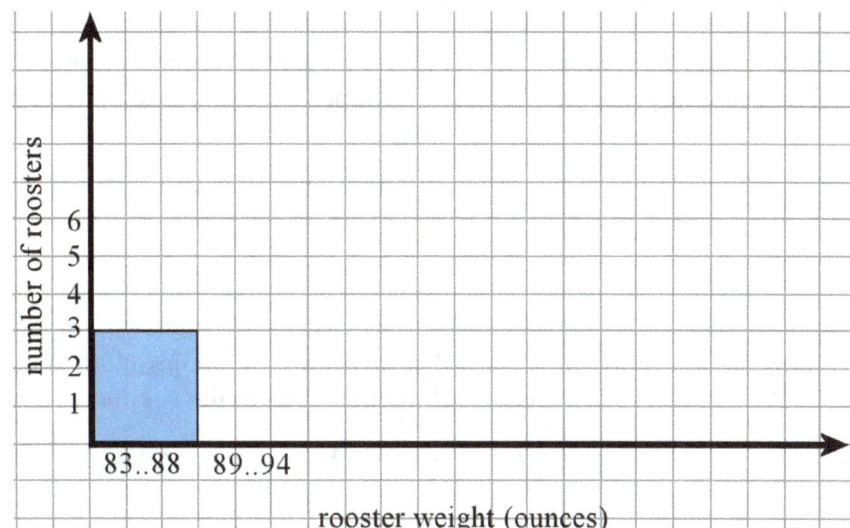

Weight (ounces)	Frequency
83..88	
89..94	
95..100	
101..106	
107..112	
113..118	

b. Maria calculated the average several times and got different results from her calculator. She must have made errors in pushing the buttons! Use the graph and the data to figure out which one is the *correct* average weight: 89 ounces, 95 1/2 ounces, or 100 1/2 ounces?

8. Solve.

a. A package of small spoons costs $13. A whole silverware set is four times as expensive. How much do *both* items cost together?

_____

_____

b. A guitar class costs $18. Ernest paid for eight classes from the $200 that he had saved. How much money does he still have?

_____

_____

c. Pete went to sleep at 22:15 and woke up at 7:00. But he also woke up at 3:30 and couldn't sleep till 5:10. How many hours and minutes did he actually sleep during the night?

Mixed Review 10

1. Choose the number sentence that fits the situation. Solve for x. Think: what does x mean in the situation?

a. Dana had \$18 and then earned more money. Now she has \$33. $\$18 + x = \$33$ OR $\$18 + \$33 = x$	**b.** The cost for dinner was \$86. Dad paid with a \$100 bill. $x - \$86 = \$100$ OR $\$100 - \$86 = x$
c. Cindy dropped ten dozen eggs and all but 39 broke. $120 - 39 = x$ OR $x - 120 = 39$	**d.** The shelter gave away 13 dogs in one day. They still had 43 dogs at the shelter. $13 + 43 = x$ OR $43 - x = 13$

2. Divide. Check each answer by multiplying.

a.	Check:	b.	Check:
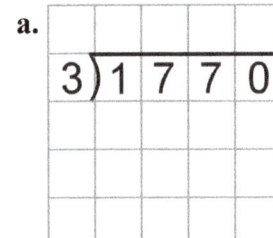		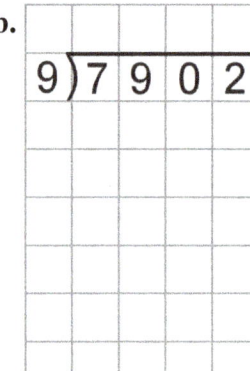	

3. Divide. Indicate the remainder.

a. $53 \div 10 =$ _____ R_____	**b.** $48 \div 9 =$ _____ R_____	**c.** $29 \div 3 =$ _____ R_____
$31 \div 11 =$ _____ R_____	$44 \div 12 =$ _____ R_____	$100 \div 9 =$ _____ R_____

4. Estimate by rounding one or both factors. Don't round both if you can calculate by just rounding one!

a. 7×78	**b.** 13×67	**c.** 311×8
$\approx$ ____ $\times$ _____ = _____	$\approx$ ____ $\times$ ____ = _____	$\approx$ _____ $\times$ ____ = _____

5. Add.

a. $60,000 + 70 =$ _____

b. $123,000 + 4,000 + 4 =$ _____

c. $3 + 90,000 + 40 =$ _____

d. $7 + 20 + 632,000 =$ _____

6. Convert between the measures of length.

a. 7 m = _____ cm	**b.** 2 m 6 cm = _____ cm	**c.** 4 km 100 m = _____ m
69 mm = ____ cm _____ mm	6 km = _____ m	169 cm = ___ m _____ cm

7. Convert between the measures of weight.

a. 3 lb 8 oz = _____ oz	**b.** 32 oz = _____ lb	**c.** 7 lb 2 oz = _____ oz
4 kg 11 g = _____ g	4,900 g = ____ kg _____ g	36 kg 140 g = _____ g

8. Solve.

a. While on vacation, the Smith family paid for different hotels this way: $145 for one night, $185 for one night, $270 for one night, and $220 for one night.
What was the average cost per night?

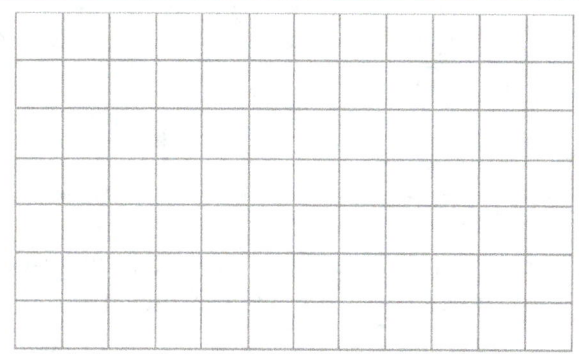

b. Isabelle walked around a square-shaped park.
Each of the sides is 650 m long.
How far did Isabelle walk?
Give your answer in kilometers and meters.

c. Andrea bought 3 notebooks for $0.86 each, a package of pencils for $0.59 and a lunchbox for $3.95. She paid with a ten-dollar bill.
How much did she spend?

What was her change?

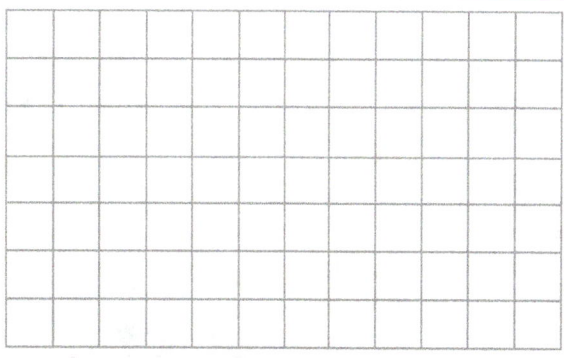

Fractions Review

1. Add or subtract. Give your final answer as a whole number or as a mixed number if possible.

a. $\dfrac{3}{5} + \dfrac{2}{5} =$	**b.** $4\dfrac{2}{8} + \dfrac{7}{8} =$	**c.** $2\dfrac{3}{4} + 4\dfrac{1}{4} =$
d. $\dfrac{9}{10} - \dfrac{7}{10} =$	**e.** $2\dfrac{1}{4} - \dfrac{3}{4} =$	**f.** $8\dfrac{9}{12} - 2\dfrac{2}{12} =$

2. Find the missing fractions.

a. $\dfrac{3}{10} + \boxed{} = 1$	**b.** $3\dfrac{2}{5} + \boxed{} = 4$	**c.** $\dfrac{4}{5} + \dfrac{3}{5} + \boxed{} = 2\dfrac{1}{5}$	**d.** $7 - \boxed{} = 6\dfrac{3}{8}$

3. Add.

a. $\dfrac{3}{10} + \dfrac{3}{100}$	**b.** $\dfrac{1}{10} + \dfrac{43}{100}$	**c.** $\dfrac{57}{100} + \dfrac{6}{10}$

4. Jane drank 1/4 liter of water from a full 2-liter pitcher.
   How much water is left in the pitcher?

5. Draw a picture showing that 2/5 = 4/10.

6. Write the equivalent fraction. Use multiplication.

a. Split all the pieces into two new ones.	**b.** Split all the pieces into five new ones.		
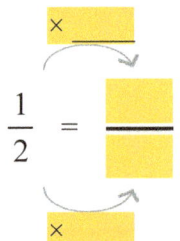 $\dfrac{3}{4} = \dfrac{\boxed{}}{\boxed{}}$	$\dfrac{1}{2} = \dfrac{\boxed{}}{\boxed{}}$	**c.** $\dfrac{2}{5} = \dfrac{4}{\boxed{}}$	**d.** $\dfrac{2}{3} = \dfrac{\boxed{}}{9}$
		e. $\dfrac{2}{3} = \dfrac{\boxed{}}{12}$	**f.** $\dfrac{3}{4} = \dfrac{12}{\boxed{}}$

7. Compare the fractions and mixed numbers.

a. $\dfrac{2}{9}$ ☐ $\dfrac{2}{10}$ b. $\dfrac{7}{10}$ ☐ $\dfrac{4}{5}$ c. $\dfrac{5}{10}$ ☐ $\dfrac{3}{6}$ d. $1\dfrac{1}{3}$ ☐ $1\dfrac{1}{5}$

e. $\dfrac{3}{4}$ ☐ $\dfrac{5}{8}$ f. $\dfrac{4}{9}$ ☐ $\dfrac{1}{3}$ g. $\dfrac{5}{12}$ ☐ $\dfrac{1}{3}$ h. $3\dfrac{4}{7}$ ☐ $3\dfrac{5}{6}$

8. Multiply.

a. $3 \times \dfrac{3}{10} =$	b. $3 \times \dfrac{2}{5} =$	c. $\dfrac{2}{10} \times 7 =$
d. $9 \times \dfrac{11}{100} =$	e. $4 \times \dfrac{5}{8} =$	f. $\dfrac{11}{12} \times 3 =$

9. Quadruple this recipe (*make it four times*).

Mexican Coffee

1 ½ cups strong gourmet coffee

¾ tsp cinnamon

4 tsp chocolate syrup

¼ tsp nutmeg

½ cup heavy cream

1 tbsp sugar

Mexican Coffee (4x)

_____ cups strong gourmet coffee

_____ tsp cinnamon

_____ tsp chocolate syrup

_____ tsp nutmeg

_____ cup heavy cream

_____ tbsp sugar

10. Remember division? Find the amounts.

a. $\dfrac{1}{5}$ of 200	b. $\dfrac{1}{6}$ of 48 cm	c. $\dfrac{2}{3}$ of 600 kg
$\dfrac{2}{5}$ of 200	$\dfrac{5}{6}$ of 48 cm	$\dfrac{5}{8}$ of $64

11. Mark got $20 for his birthday. The same day he spent 3/4 of it.
Now how much does he have left?

12. Rose has read 3/8 of a 240-page book.
How many pages are still left to read?

Fractions Test

1. Add and subtract.

a. $\dfrac{3}{6} + \dfrac{2}{6} + \dfrac{1}{6} =$	**b.** $1\dfrac{2}{3} + \dfrac{2}{3} =$	**c.** $3\dfrac{3}{5} + 2\dfrac{1}{5} =$
d. $\dfrac{11}{12} - \dfrac{7}{12} - \dfrac{2}{12} =$	**e.** $3\dfrac{1}{5} - \dfrac{3}{5} =$	**f.** $7\dfrac{5}{6} - 2\dfrac{1}{6} =$

2. Arrange the fractions in order from the smallest to the greatest.

a. $\dfrac{3}{4}$, $\dfrac{3}{8}$, $\dfrac{1}{2}$	**b.** $\dfrac{5}{5}$, $\dfrac{5}{7}$, $\dfrac{7}{5}$	**c.** $\dfrac{5}{9}$, $\dfrac{5}{2}$, $\dfrac{5}{6}$

3. Split both the colored and white pieces as asked. Write the fraction before and the fraction after.

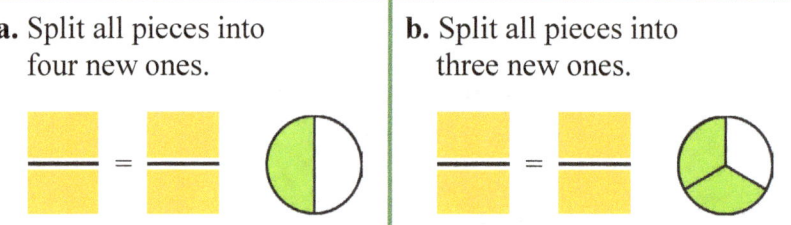

a. Split all pieces into four new ones.

b. Split all pieces into three new ones.

4. Write the equivalent fraction. Use multiplication.

a. $\dfrac{1}{5} = \dfrac{2}{}$ **b.** $\dfrac{3}{4} = \dfrac{}{12}$ **c.** $\dfrac{4}{5} = \dfrac{}{25}$ **d.** $\dfrac{1}{6} = \dfrac{4}{}$

5. Multiply. Give your answer as a whole number or as a mixed number.

a. $3 \times \dfrac{4}{10} =$	**b.** $5 \times \dfrac{3}{5} =$	**c.** $\dfrac{3}{8} \times 4 =$

6. Three brothers made a big pizza and divided it into 12 equal pieces. Walter ate 1/4 of the pizza, John ate 1/12 of it, and Eric ate three times as much as John.

   a. Who ate the most pizza?

   b. How much more did Eric eat than John?

Mixed Review 11

1. Solve the division in your head. Then write a multiplication and addition sentence that checks your division.

a. $57 \div 5 =$ _____	**b.** $34 \div 7 =$ _____	**c.** $33 \div 9 =$ _____
____ × ____ + ____ = ____	____ × ____ + ____ = ____	____ × ____ + ____ = ____

2. Find the areas of these rectangles. Remember to use the right unit for the area!

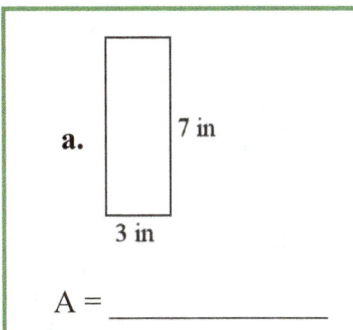

a. 7 in, 3 in

A = _____

b. 20 km, 25 km

A = _____

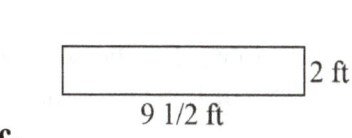

c. 2 ft, 9 1/2 ft

A = _____

3. Identify these triangles as right, obtuse, or acute triangles.

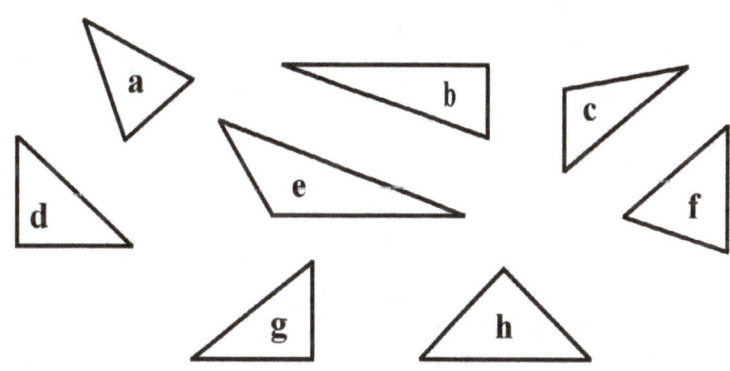

4. **a.** Draw a parallelogram by drawing two lines that intersect the two given lines.

   b. Measure all the sides of the parallelogram and write their lengths in the figure, next to each side.

   c. Find the angle measures of your parallelogram.

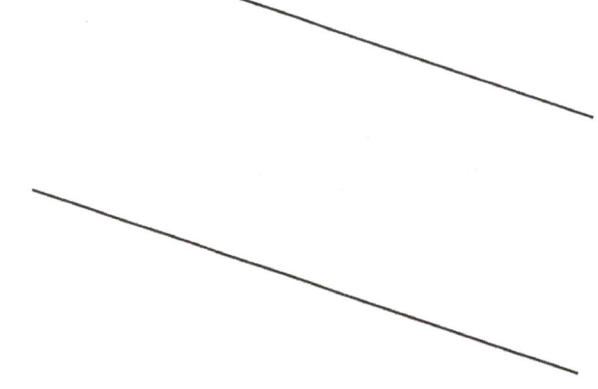

5. Divide. Check your answers.

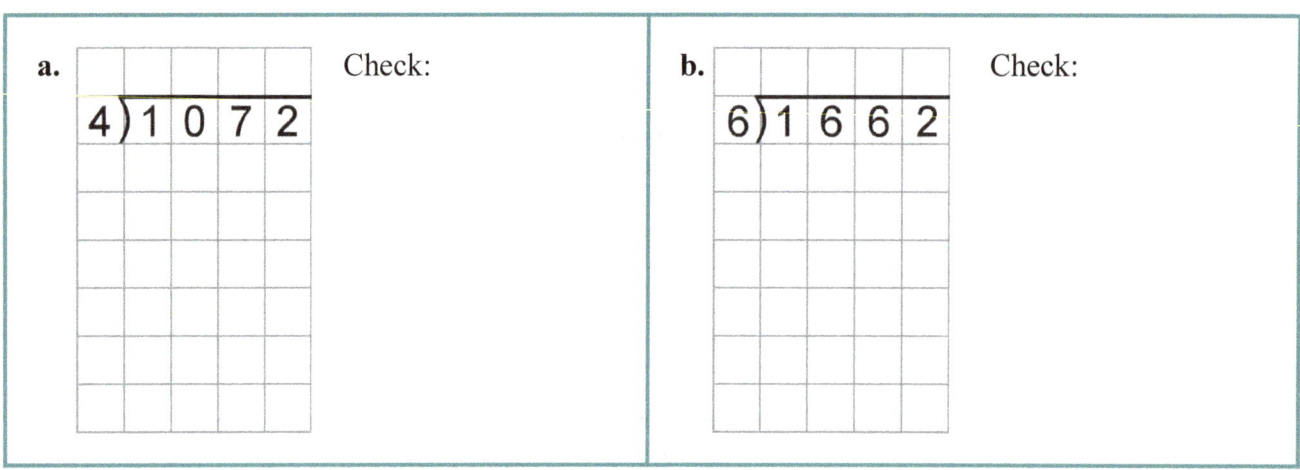

a.   4)1 0 7 2   Check:

b.   6)1 6 6 2   Check:

6. Solve.

a. Which weighs more, four boxes that are
44 pounds each, or five boxes that
are 32 pounds each?

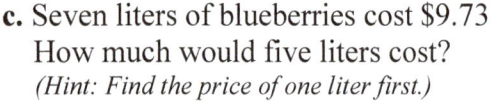

_____

_____

How much more?

_____

b. A total of 86 balloons were divided evenly among
25 children, and the teacher kept the remainder.
How many did the teacher get?

_____

c. Seven liters of blueberries cost $9.73.
How much would five liters cost?
(Hint: Find the price of one liter first.)

7. Write a single number sentence that tells you
the change if you buy a book for $7, a ball for $5,
and pay with a $20 bill.

Mixed Review 12

1. Estimate by rounding the numbers to the nearest thousand or to the nearest ten thousand. Then calculate.

a. $22,934 + 5,312 + 424,787$	**b.** $519,313 - 47,616$
Estimation:	Estimation:

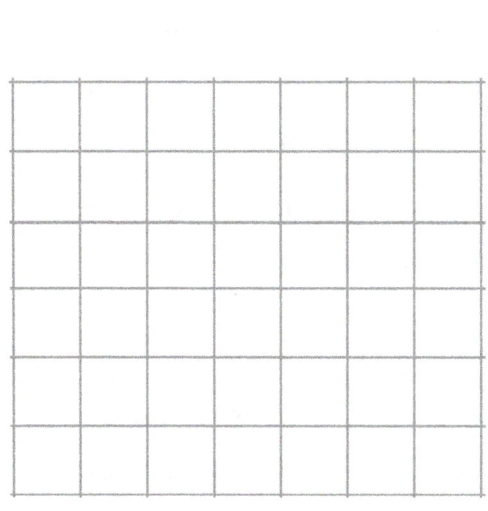
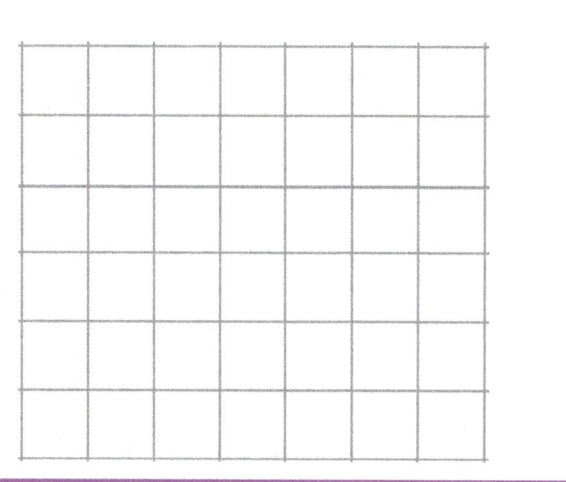

2. There are 45 students in each of the 22 buses, and 27 students in one additional bus. What was the total number students in the 23 buses?

3. Fill in. Then draw an example picture of each triangle.

Right angles are exactly _____°.

Right triangles have exactly _____ right angle.

Obtuse angles are more than _____° but less than _____°.

Obtuse triangles have exactly _____ obtuse angle.

Acute angles are less than _____°.

Acute triangles have _____ acute angles.

4. If the perimeter of a rectangle is 28 inches, then what could the side lengths be? Write possible side lengths in the table. Then calculate the areas. You can draw the rectangles in the grid, if you wish.

One side	Other side	Perimeter	Area
		28 in	
		28 in	
		28 in	
		28 in	

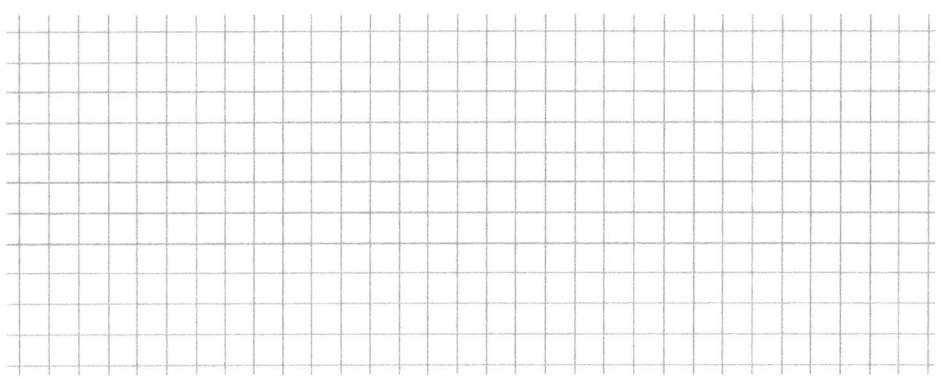

5. Convert the time to the 24-hour clock time.

a. 1:40 p.m.	**b.** 9:20 p.m.	**c.** 2:15 p.m.	**d.** 10:04 a.m.
_____ : _____	_____ : _____	_____ : _____	_____ : _____

6. Mark an "x" if the number is divisible by 2, by 5, or by 10.

number	divisible			number	divisible			number	divisible		
	by 2	by 5	by 10		by 2	by 5	by 10		by 2	by 5	by 10
478				1,492				904			
540				3,093				905			
255				94				906			

7. Is 549 divisible by 7?
   Explain why or why not.

8. Find all the factors of the given numbers. Think of writing the number as a multiplication in many
   different ways. Don't forget 1 as a factor!

a. 24	b. 66
factors:	factors:
c. 96	d. 75
factors:	factors:

9. Draw the liquid in the thermometer. Match the temperatures to show the situations.

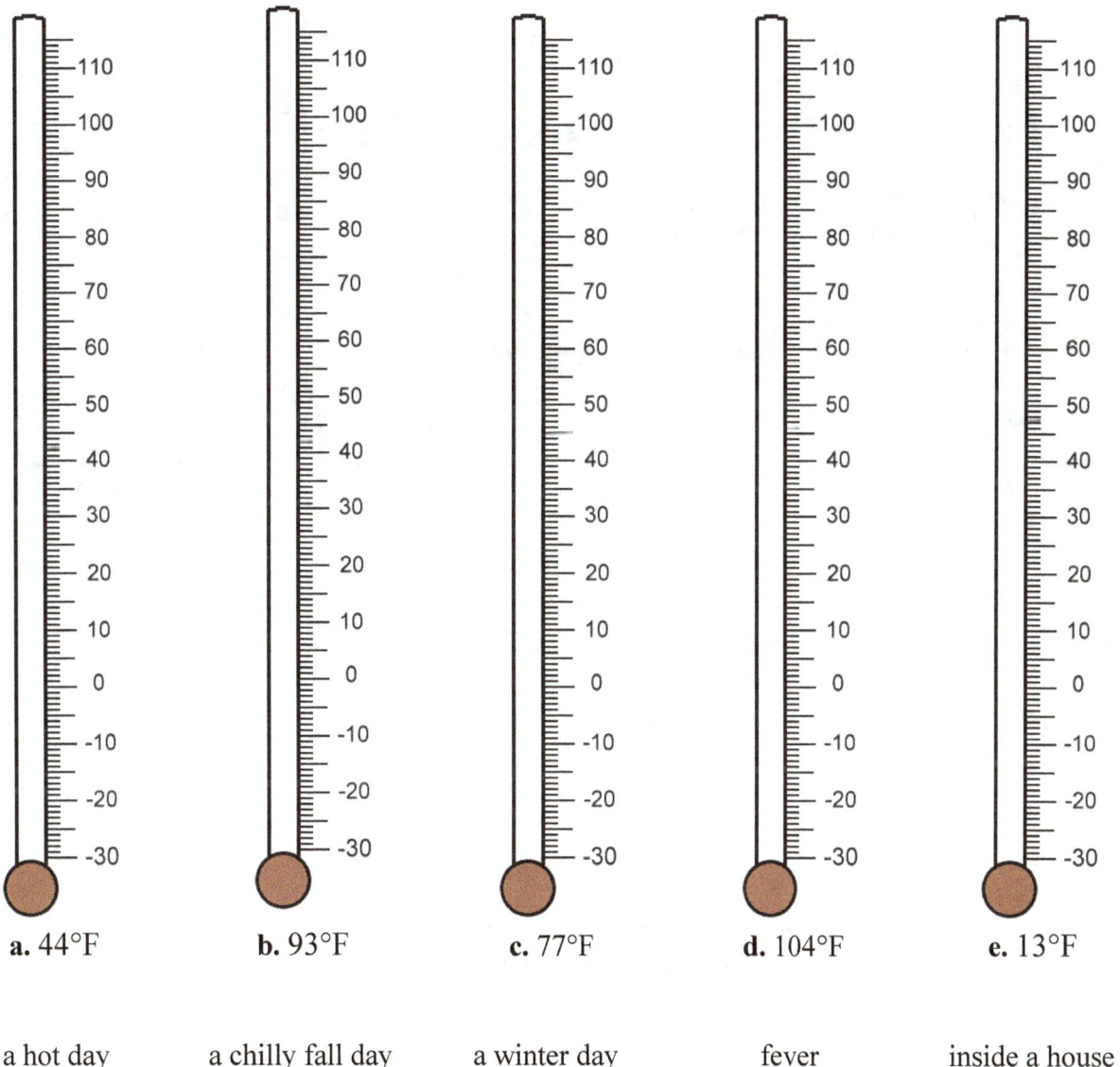

a. 44°F   b. 93°F   c. 77°F   d. 104°F   e. 13°F

a hot day   a chilly fall day   a winter day   fever   inside a house

Decimals Review

1. Write the fractions as decimals and vice versa.

a. $\frac{7}{10}$	**b.** $\frac{7}{100}$	**c.** $1\frac{6}{10}$	**d.** $2\frac{41}{100}$
e. $1\frac{1}{100}$	**f.** $\frac{47}{100}$	**g.** 0.8	**h.** 2.9
i. 4.14	**j.** 18.08	**k.** 0.03	**l.** 0.29

2. Compare. Write < , > , or = between the numbers. Use the place value charts to help.

a. 0.5 ⬚ 0.05   **b.** 0.3 ⬚ 0.28   **c.** 2.15 ⬚ 2.2

d. 4.50 ⬚ $4\frac{1}{2}$ **e.** 4.87 ⬚ 4.78   **f.** 2.30 ⬚ 2.3

g. 0.98 ⬚ 1.01   **h.** 3.77 ⬚ 7.37   **i.** 2.7 ⬚ $2\frac{1}{2}$

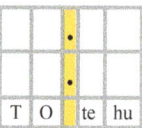
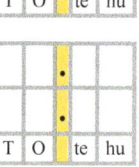

3. Mark the following decimals on the number line: 0.47   0.60   0.09   0.91   0.02   0.38

0	0.1	0.2	0.3	0.4	0.5	0.6	0.7	0.8	0.9	1

4. Write in order from the smallest to the largest number:

0.1   0.21   0.12   0.2 $\frac{1}{2}$ 0.8   0.74

5. Add or subtract. Write each problem using fractions also. To help you, tag a zero to the shorter decimal so that both decimals have two decimal digits.

a. 0.7 + 0.03 =	**b.** 0.32 + 0.4 = _____	**c.** 0.7 − 0.04 = _____
$\frac{}{100} + \frac{}{100} = \frac{}{100}$	$\frac{}{100} + \frac{}{100} = \frac{}{100}$	$\frac{}{100} - \frac{}{100} = \frac{}{100}$

6. Add or subtract in your head.

a. 0.3 + 0.7 = _____

b. 0.12 + 0.76 = _____

c. 0.3 + 0.06 = _____

d. 0.3 − 0.06 = _____

e. 0.13 + 0.7 = _____

f. 1.3 − 0.8 = _____

7. Some of these additions are wrong. Be a teacher detective, and correct the ones that are wrong.

a. 0.99 + 0.1 = 1	**b.** 0.3 + 0.05 = 0.35
c. 0.19 + 0.19 = 1.38	**d.** 0.03 + 0.5 = 0.08

8. Add or subtract in columns. Line up the decimal points.

a. 2.7 + 6.61	**b.** 15 + 7.21 + 0.9	**c.** 8.2 − 2.36

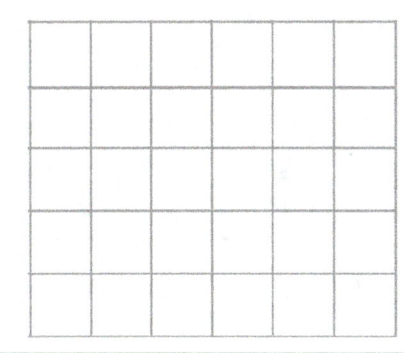
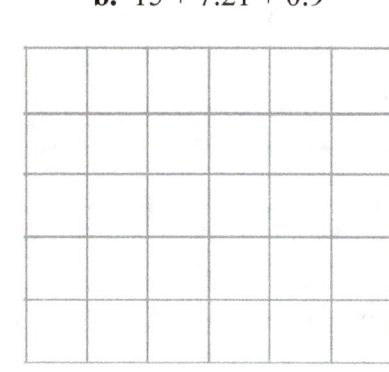
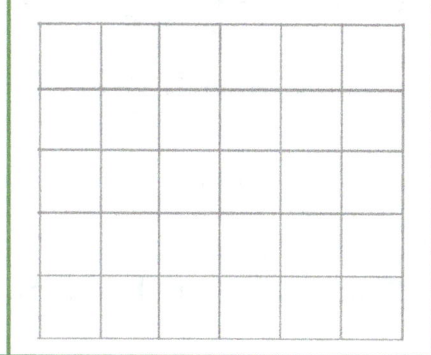

9. Calculate.

40 − (22.46 + 14.7)

10. Remember? One kilogram is 1,000 grams.

How many grams is 9/10 of a kilogram?

How many grams is 0.2 kg?

Which is heavier, a tablet computer that weighs 610 grams, or one that weighs 0.6 kg?

Decimals Test

1. Mark these decimals on the number line: 1.60   1.21   1.78   1.04

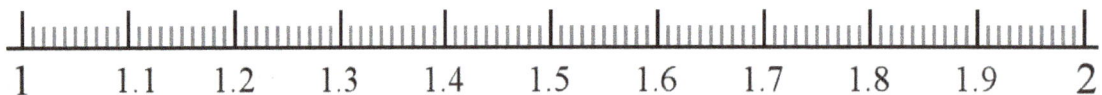

2. Write the fractions as decimals and decimals as fractions.

a. $\dfrac{2}{10}$	**b.** $7\dfrac{4}{100}$	**c.** $\dfrac{74}{100}$	**d.** 0.52	**e.** 3.9

3. Add and subtract.

a. $0.5 + 1.7 =$ _____

b. $0.44 + 0.51 =$ _____

c. $0.2 - 0.01 =$ _____

d. $1.6 - 0.9 =$ _____

e. $0.3 + 0.07 =$ _____

f. $5.05 - 2.01 =$ _____

4. Compare. Write $<$, $>$, or $=$ between the numbers.

a. 0.4 ☐ 0.14	**b.** 2.9 ☐ 2.90	**c.** 4.3 ☐ 4.03	**d.** 0.45 ☐ $\dfrac{1}{2}$	**e.** 7.18 ☐ 7.8

5. Write in order from the smallest to the greatest number: 7.2   2.7   2.07   2.17   2.77

6. Find the total weight of four books that weigh 1.3 kg each.

7. Calculate.

a. $4.56 + 2.8$	**b.** $4.56 - 2.8$
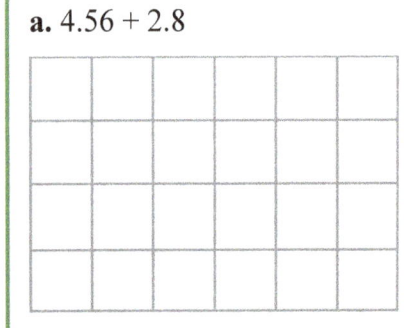	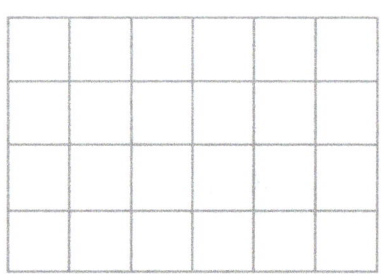

Mixed Review 13

1. Write an addition sentence. Give your answer as a mixed number.

a.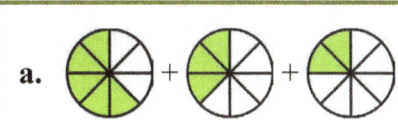

b.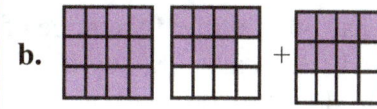

2. Multiply.

a. $10 \times \dfrac{5}{12}$	b. $\dfrac{4}{9} \times 7$	c. $\dfrac{14}{100} \times 3$

3. Find all the factors of the numbers.

a. 38	b. 56	c. 19

4. Multiply. First, estimate the answer on the empty line.

a. 6×292	b. 11×402	c. $3 \times 2,364$	d. $7 \times 8,827$
≈ _____	≈ _____	≈ _____	≈ _____

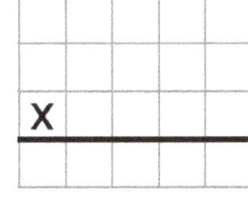 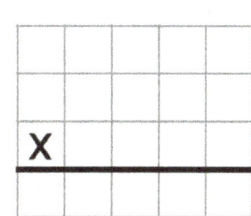 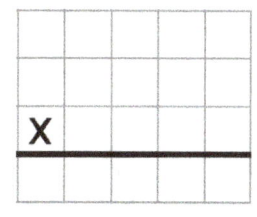 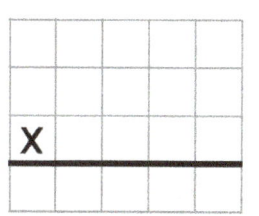

5. Solve the equations.

a. $90 \times \triangle = 8,100$ $\triangle =$ _____	b. $500 \times \underline{?} = 2 \times 1,000$ $\underline{?} =$ _____	c. $4 \times 3 \times y = 360$ $y =$ _____

6. Change the 24-hour times to the a.m. / p.m. times.

a. 14:30	**b.** 19:15	**c.** 22:45	**d.** 7:50
_____ : _____	_____ : _____	_____ : _____	_____ : _____

7. Mom promised to pay one-fourth of the price of a new
   $95 bicycle for Terry. Draw a bar model for the situation,
   and find how much Mom and Terry each paid.

8. Jack has paid 3/5 of the $600-computer he bought.
   How many dollars does he still have left to pay?

9. Draw as many different symmetry lines as you can into this shape.

10. **a.** Draw a 35° angle using B as vertex and AB as one side of the angle. If you do it right, the other
    side of your angle will intersect (cross) the line segment AC so that you will get a triangle.

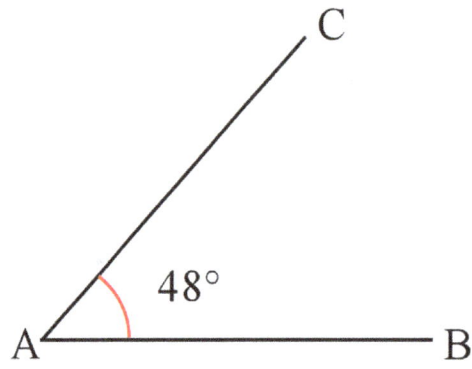

   b. Measure the third angle of the triangle.

Mixed Review 14

1. Multiply each part of the number (ones, tens, and hundreds) separately, then add.

a. 4×36	**b.** 5×65	**c.** 8×426
_____ + _____	_____ + _____	_____ + _____ + _____
= _____	= _____	= _____

2. If you add one thousand, one hundred, one ten, and one to this number, it becomes 100,000. What is the number?

3. Estimate the products by rounding one or both factors.

a. $8 \times 69 \approx$	**b.** $11 \times 55 \approx$	**c.** $25 \times 17 \approx$

4. Write the division problem. Solve for x.

a. The divisor is 8, the dividend is x, and the quotient is 7.   ____ ÷ ____ = ____ $x =$ ____

b. The dividend is 24, the divisor is x, and the quotient is 8.   ____ ÷ ____ = ____ $x =$ ____

5. Are the lines drawn in the shapes symmetry lines for them?

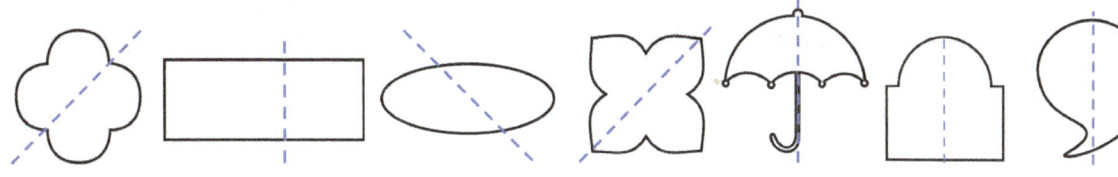

6. Convert.

a. 5 ft = _____ in	**b.** 3 ft 4 in = _____ in	**c.** 4 yd = _____ ft
12 ft = _____ in	6 ft 6 in = _____ in	9 yd = _____ ft

7. Underline the heaviest amount.

a. 5 kg 500 g 5,050 g	**b.** 340 g 3 kg 400 g	**c.** 9 kg 9,900 g 900 g

8. Solve the problems.

   a. Jack weighed the new kittens when they were two weeks old.
      They weighed 3 oz, 3 oz, 5 oz, 2 oz, and 4 oz.
      What was their total weight in pounds and ounces?

   b. Annie packed 175 kg of strawberries into 4-kg boxes.
      How many boxes did she need?

   c. Ava had $98 saved. She used 1/7 of her savings to buy balloons
      and other items for a party. How much money is left of her savings?

   d. Thirty-six children are going to march in rows in a parade.
      How many children should be in each row so that
      the rows will be even?

   e. Two-thirds of the horses on a farm are adults and
      the rest are foals. There are 68 adult horses.

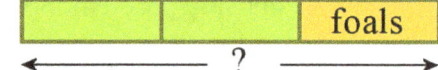

      How many foals are there?

      How many horses total are there?

9. Draw angles of the following measures. Use a protractor.

a. 63°	**b.** 108°

10. Find rays, lines, and line segments that are either parallel or perpendicular to each other. Use these notations: ∥ for parallel and ⊥ for perpendicular.

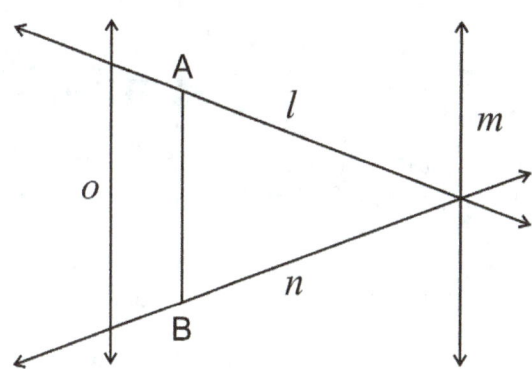

11. An apple is cut into 8 pieces. You ate 2/8 of it. How many fourths of the apple are left?

12. Add and subtract. Give your answer as a mixed number or a whole number.

a. $1\dfrac{2}{7} + 1\dfrac{5}{7} =$	**b.** $\dfrac{4}{12} + \dfrac{9}{12} =$
c. $\dfrac{1}{10} + \dfrac{3}{100} =$	**d.** $5 - 3\dfrac{3}{4} =$
e. $\dfrac{46}{100} - \dfrac{2}{10} =$	**f.** $2\dfrac{4}{10} + 3\dfrac{7}{10} =$

13. Multiply. Give your answer as a mixed number or a whole number.

a. $6 \times \dfrac{5}{10}$	**b.** $2 \times \dfrac{3}{5}$	**c.** $5 \times \dfrac{3}{4}$	**d.** $\dfrac{7}{10} \times 10$

14. Write the equivalent fractions. Shade parts in the pictures.

a. $\dfrac{4}{5} =$	**b.** $\dfrac{1}{3} =$	**c.** $\dfrac{2}{3} =$

15. Compare. Write < or > in between the fractions.

a. $\dfrac{3}{8} \square \dfrac{2}{4}$	**b.** $\dfrac{6}{8} \square \dfrac{8}{5}$	**c.** $\dfrac{2}{5} \square \dfrac{4}{7}$	**d.** $\dfrac{5}{10} \square \dfrac{5}{12}$

End-of-Year Test Grade 4

This test is quite long, so I do not recommend that your child/student does all of it in one sitting. Break it into parts and administer them on several days. Use your judgment.

This is to be used as a diagnostic test. Thus, you may even skip those areas and concepts that you already know for sure your student has mastered.

The test does not cover every single concept that is covered in the *Math Mammoth Grade 4 Complete Curriculum,* but all the major concepts and ideas are tested here. This test is evaluating the child's ability in the following content areas:

- addition and subtraction
- early algebraic thinking
- the order of operations
- graphs
- large numbers and place value
- rounding and estimating
- multi-digit multiplication
- word problems
- some basic conversions between measuring units
- measuring length
- time calculations
- long division
- the concept of remainder
- factors
- area and perimeter
- measuring and drawing angles
- classifying triangles according to their angles
- adding and subtracting fractions and mixed numbers (like fractional parts)
- equivalent fractions
- comparing fractions
- multiplying fractions by whole numbers
- the concept of a decimal (tenths/hundredths)
- comparing decimals

In order to continue with the Math Mammoth Grade 5 Complete Curriculum, I recommend that the child gain a minimum score of 80% on this test, and that the teacher or parent review with him any content areas where he is weak. Children scoring between 70 and 80% may also continue with grade 5, depending on the types of errors (careless errors or not remembering something, vs. the lack of understanding). The most important content areas to master are multi-digit multiplication, long division, place value, and word problems. Again, use your judgment.

Instructions

A calculator is not allowed. My suggestion for grading is below. The total is 192 points. A score of 154 points is 80%.

Question	Max. points	Student score
Addition, Subtraction, Patterns, and Graphs		
1	2 points	
2a	1 point	
2b	2 points	
3	2 points	
4	6 points	
5	4 points	
6	2 points	
7	4 points	
8	3 points	
	subtotal	/ 26
Large Numbers and Place Value		
9	3 points	
10	2 points	
11	3 points	
12	3 points	
13	2 points	
14	3 points	
15	3 points	
16	4 points	
	subtotal	/ 23
Multi-Digit Multiplication		
17	6 points	
18	3 points	
19	8 points	
20	3 points	
21a	3 points	
21b	2 points	
21c	2 points	
21d	3 points	
	subtotal	/ 30

Question	Max. points	Student score
Time and Measuring		
22	2 points	
23	1 point	
24	3 points	
25	2 points	
26	6 points	
27	6 points	
28	2 points	
29	1 point	
30	2 points	
	subtotal	/ 25
Division and Factors		
31	4 points	
32	3 points	
33	4 points	
34a	3 points	
34b	2 points	
35	6 points	
36	4 points	
37	2 points	
38	4 points	
	subtotal	/ 32
Geometry		
39	2 points	
40	2 points	
41	3 points	
42	2 points	
43	2 points	
44	1 point	
45	2 points	
46	3 points	
	subtotal	/ 17

Question	Max. points	Student score
Fractions and Decimals		
47	1 point	
48	1 point	
49	3 points	
50	2 points	
51	4 points	
52	4 points	
53	2 points	
54	1 point	
55	3 points	
56	4 points	
57	4 points	
58	4 points	
59	4 points	
60	2 points	
	subtotal	/ 39
	TOTAL	/ 192

End-of-Year Test - Grade 4

Addition, Subtraction, Patterns, and Graphs

1. Subtract. Check by adding.

$5{,}200 - 2{,}677 - 543$	Add to check:

2. **a.** First round the prices to the nearest dollar. Then use the rounded prices to estimate the total bill.

Crackers $1.28; Cheese $8.92; Jam $3.77; Butter $9.34.

b. Now, use the exact prices (not rounded prices). Mrs. Wood buys the items listed above and pays with $30. What is her change?

3. *Estimate* the cost of buying five notebooks for
   $0.87 each and two pencil cases for $1.24 each.

4. Calculate in the right order.

a. $3 \times (4 + 6) =$ _____	**b.** $3 \times 3 + 8 \div 4 =$ _____	**c.** $20 \times 3 + 80 \div 1 =$ _____
$100 - 4 \times 4 =$ _____	$(7 - 3) \times 3 + 2 =$ _____	$15 + 2 \times (8 - 6) =$ _____

5. Circle the number sentence that fits the problem. Then solve for *x*.

a. Alice had $35. Then she earned more money (*x*). Now she has $92.	**b.** Eric gave 24 of the cookies he had baked to a friend and now he has 37 cookies left.
$\$35 + x = \$92$ OR $\$35 + \$92 = x$	$37 - 24 = x$ OR $x - 24 = 37$
$x =$ _____	$x =$ _____

6. **a.** Continue this pattern for four more numbers:

   2,000   1,750   1,500   1,250

   b. Write a list of six numbers that follows this pattern: Start at 200, and add 300 each time.

7. These numbers are the students' quiz scores: 2 5 8 7 6 6 7 10 10 4 7 7 8 6 8 5 9 9 8 6 6 5 7 9
   Make a frequency table and a bar graph.

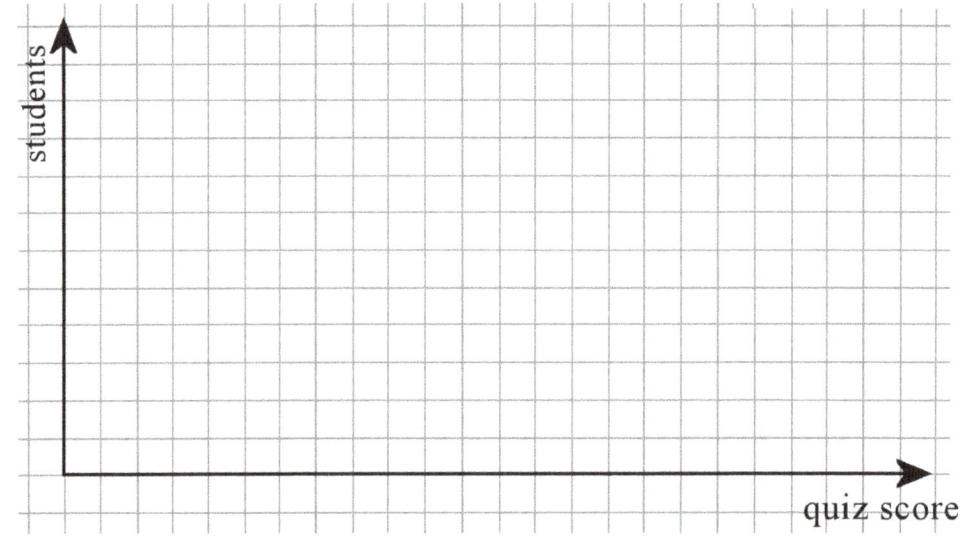

Quiz score	Frequency
1	
2	
3	
4	
5	
6	
7	
8	
9	
10	

students

quiz score

8. Write an addition or a subtraction with an unknown (*x* or ?). Solve it. The bar model can help.

Rubber boots used to cost $27.95 but now the price is $21.45. How much is the discount?

←—— original price ——→

_____

_____

Large Numbers and Place Value

9. Subtract in your head.

a. $2,000 - 1 =$ _____	**b.** $5,000 - 20 =$ _____	**c.** $6,000 - 300 =$ _____

10. Write the numbers in the normal form.

a. 800 thousand 50

b. 25 thousand 4 hundred 7

11. Find the missing numbers.

a. $30,550 = 50 +$ _____ $+ 500$	**b.** $809,100 = 800,000 + 100 +$ _____
c. $725,608 = 20,000 + 700,000 + 8 +$ _____ $+ 5,000$	

12. Compare, writing $<$, $>$, or $=$ between the numbers.

a. 54,500 55,400	**b.** 108,882 108,828	**c.** 71,600 61,700

13. Write the numbers in order from the smallest to the greatest.
217,200   227,712   27,200   227,200

14. Round the numbers as the dashed line indicates (to the underlined digit).

a. $43\underline{6},102 \approx$ **b.** $8\underline{9},756 \approx$ **c.** $27,\underline{5}29 \approx$

15. Round to the nearest ten thousand.

a. $426,889 \approx$ **b.** $495,304 \approx$ **c.** $7,345 \approx$

16. Calculate. Line up the digits that are in the same place carefully.

    a. 476,708 + 24,392 + 563

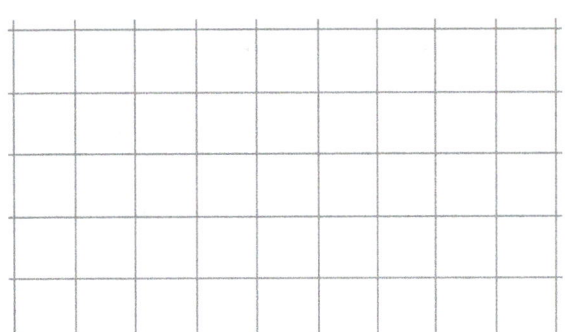

    b. 405,112 − 81,424

Multi-Digit Multiplication

17. Multiply, and find the missing factors.

a. 70 × 3 = _____	**b.** 6 × 800 = _____	**c.** 40 × 80 = _____
d. _____ × 3 = 360	**e.** 50 × _____ = 4,000	**f.** _____ × 300 = 21,000

18. Ed earns $20 per hour.

    a. How much will he earn in an 8-hour workday? _____

    b. How much will he earn in a 40-hour workweek? _____

    c. How many days will he need to work in order to earn at least $600? _____

19. Multiply. Estimate the answer on the line.

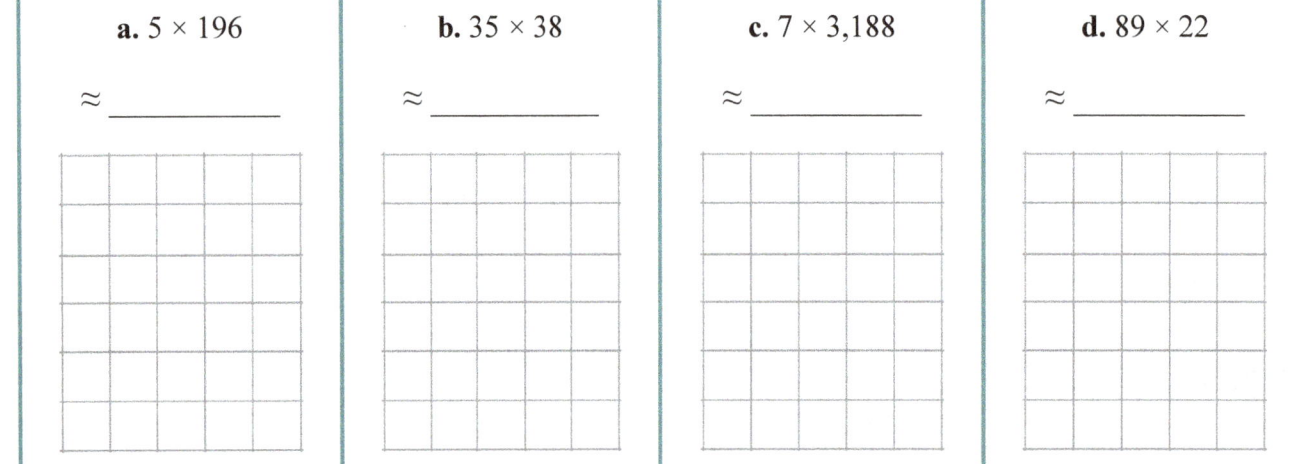

a. 5 × 196	**b.** 35 × 38	**c.** 7 × 3,188	**d.** 89 × 22
≈ _____	≈ _____	≈ _____	≈ _____

20. Write the area of the *whole* rectangle as a SUM of the areas of the smaller rectangles. Lastly, add
to find the total area.

Area = 8 × 127

= ___ × _____ + ___ × ___ + ___ × ___

=

21. Solve. Don't give just the answer but also <u>write a number sentence</u> or several for each problem.

a. Find the change, if Sally buys 26 shirts
for $14 each, and pays with $400.

_____

_____

Estimate: _____

b. How many minutes are there in a day (24 hours)?

_____

_____

c. One side of a square is 375 cm.
What is its perimeter?

_____

_____

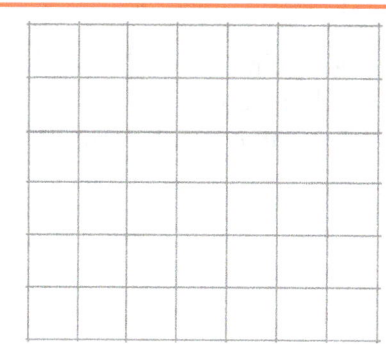

d. Bicycles that cost $277 were discounted by $58.
A store bought eight. What was the total cost?

_____

_____

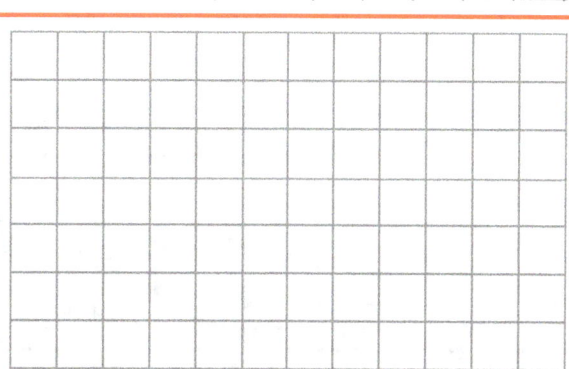

Time and Measuring

22. Measure the lines below to the nearest eighth of an inch and also in centimeters and millimeters.

_____

a. _____ in. or _____ cm _____ mm

_____

b. _____ in. or _____ cm _____ mm

23. How much time passes from 10:54 a.m. till 5:06 p.m.?

24. Luis kept track of how long it took him to do his homework:

Monday	Tuesday	Wednesday	Thursday	Sunday
1 h 45 min	50 min	1 h 15 min	2 h 15 min	55 min

How much time did he spend with homework in total?

_____

25. A teacher started her workday at 7:00 am, and stopped it at 3:35 pm. But in between, she had a 45-minute lunch break, and another break of 20 minutes. How many hours/minutes did she actually work?

26. Convert between the different measuring units.

a.	b.	c.
6 lb = _____ oz	5 gal = _____ qt	4 ft 2 in = _____ in
2 lb 11 oz = _____ oz	2 qt = _____ cups	7 yd = _____ ft

27. Convert between the different measuring units.

a.	b.	c.
2 kg = _____ g	5 L 200 ml = _____ ml	8 cm 2 mm = _____ mm
11 kg 600 g = _____ g	3 m = _____ cm	10 km = _____ m

28. George jogs daily on a track through the woods that is 3 km 800 m long.
    What is the total distance he runs in four days?

29. Alice drank 350 ml of a 2-liter bottle of water.
    How much is left?

30. The long sides of a rectangle measure 5 ft 6 in,
    and the short sides are 3 ft 4 in.

    What is the perimeter? _____ ft _____ in

Division and Factors

31. Divide. Check each problem by multiplying.

a. 567 ÷ 9   Check:

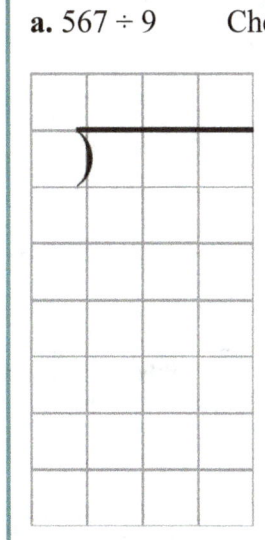

b. 8,564 ÷ 4   Check:

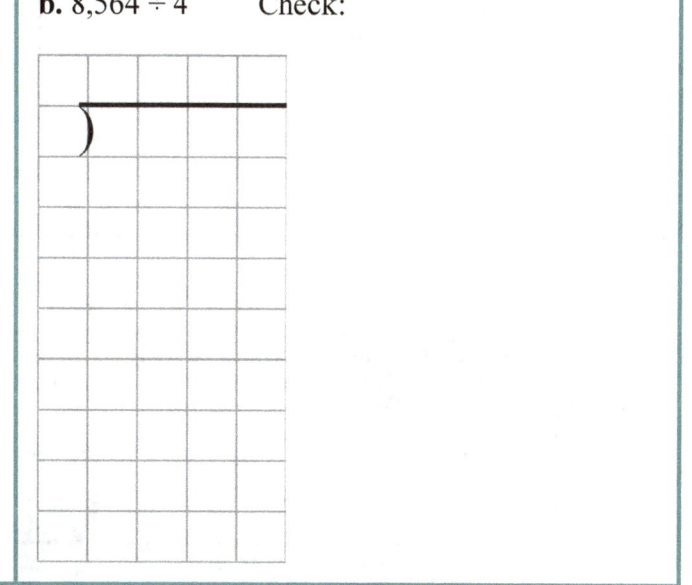

32. Solve.

| **a.** 47 ÷ 5 = _____ R ___ | **b.** 25 ÷ 3 = _____ R ___ | **c.** 57 ÷ 9 = _____ R ___ |

33. Solve.

a. Amy put 48 photographs into an online photo album.
On each page she could fit nine photos.
How many photos were on the last page?

How many pages were full?

b. You bought a 50-foot roll of chain-link fence that cost $150.
Then you sold 12 feet of it to your neighbor.
How much did your neighbor pay?

34. Solve.

a. Joe had saved $264. He spent 3/8 of that to buy a camera. How much did the camera cost?

b. Mary packed 117 muffins into bags of six. How many bags does Mary need for them?

35. Mark an X if the number is divisible by the given numbers.

number	divisible by 1	divisible by 2	divisible by 3	divisible by 4	divisible by 5	divisible by 6	divisible by 7	divisible by 8	divisible by 9	divisible by 10
80										
75										
47										

36. Answer each question with a "yes" or "no," and give a reason.

a. Is 5 a factor of 60? _____ , because _____ × _____ = _____ .	**b.** Is 7 a divisor of 43? _____ , because _____ ÷ _____ = _____ .
c. Is 96 divisible by 4? _____ , because _____ .	**d.** Is 34 a multiple of 7? _____ , because _____ .

37. List three prime numbers.

38. Find all the factors of the given numbers.

a. 56 factors:	**b.** 78 factors:

Geometry

39. Measure this angle.

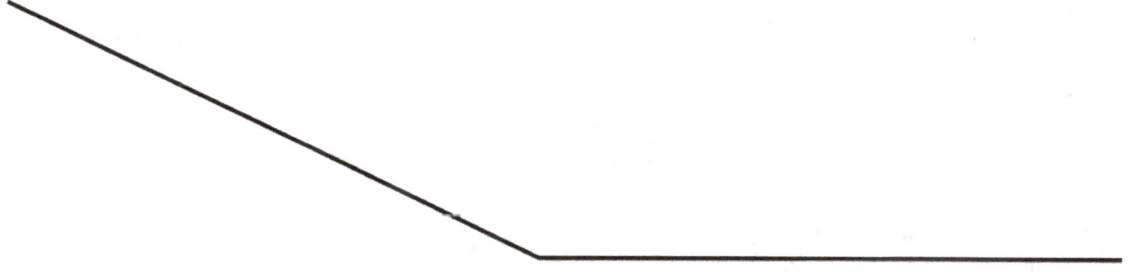

40. Draw here an angle of 65°.

41. Draw here any obtuse triangle,
    and measure its angles.

42. Write an addition sentence about
    the angle measures. Use an unknown (*x*)
    for one angle measure.

    Then solve it.

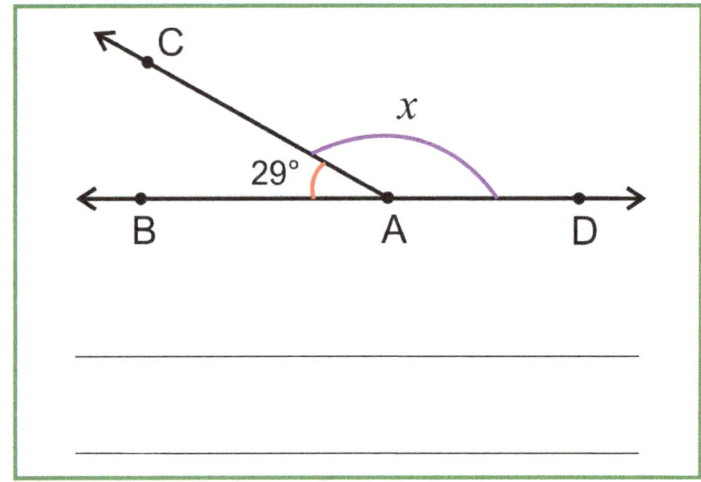

_____

_____

43. Sketch here any rectangle. Then draw a diagonal
    line in it (a line from corner to corner).
    What kind of triangles are formed?

44. Sketch here two line segments that
    are perpendicular to each other.

45. Draw as many different symmetry lines as you can into these shapes.

46. This picture shows the floor of a room with a carpet on the floor. The room itself measures 28 feet by 12 feet. The carpet is 6 ft by 10 ft. Find the area of floor outside the carpet (not including the carpet).

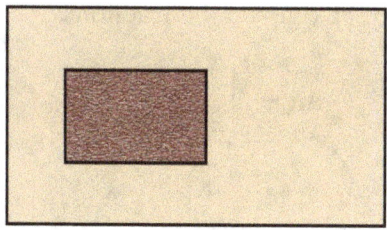

Fractions and Decimals

47. Write an addition to match the picture: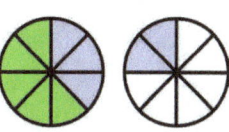

48. Erica did 1/4 of a puzzle and Mom did another fourth of it. How much of the puzzle is still left to do?

49. Add and subtract. Give your final answer as a whole number or as a mixed number if possible.

| **a.** $\dfrac{4}{5} + \dfrac{3}{5} =$ | **b.** $1\dfrac{1}{6} - \dfrac{2}{6} =$ | **c.** $3\dfrac{6}{8} + 2\dfrac{2}{8} =$ |

50. Split the existing pieces. Fill in the missing parts.

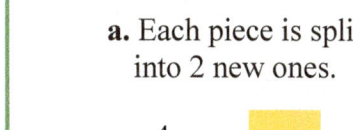

a. Each piece is split into 2 new ones.

$$\frac{4}{5} = \boxed{}$$

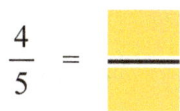

b. Each piece is split into ____ new ones.

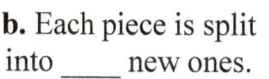

$$\boxed{} = \frac{6}{9}$$

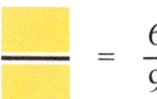

51. Write the equivalent fractions.

a. $\frac{2}{3} = \frac{}{15}$	b. $\frac{3}{5} = \frac{9}{}$	c. $\frac{1}{6} = \frac{}{12}$	d. $\frac{1}{3} = \frac{}{9}$

52. Compare the fractions.

a. $\frac{2}{3} \square \frac{3}{8}$ b. $\frac{6}{5} \square \frac{7}{8}$ c. $\frac{11}{12} \square \frac{11}{10}$ d. $\frac{1}{3} \square \frac{5}{12}$

53. Write these fractions in order, from the smallest to the greatest: $\frac{5}{4}, \frac{7}{10}, \frac{65}{100}$

54. A recipe calls for 3/4 cup of flour. If you triple the recipe, how much flour do you need?

55. Fill in.

a. $\frac{3}{8} = 3 \times \frac{}{}$	b. $4 \times \frac{2}{5} =$	c. $7 \times \frac{2}{12} =$

56. Mark on the number line the following decimals: 0.55 0.08 0.27 0.80

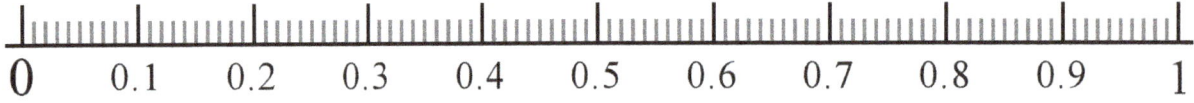

0   0.1   0.2   0.3   0.4   0.5   0.6   0.7   0.8   0.9   1

57. Write the fractions and mixed numbers as decimals.

a. $\frac{3}{10}$	b. $3\frac{9}{10}$	c. $\frac{9}{100}$	d. $7\frac{45}{100}$

58. Write the decimals as fractions or mixed numbers.

a. 0.6	b. 6.7	c. 0.21	d. 5.05

59. Compare.

a. 0.17 ☐ 0.2   **b.** 1.6 ☐ 1.56   **c.** 13.09 ☐ 13.9   **d.** 9.80 ☐ 9.8

60. Add and subtract.

a. 7.81 + 5.2	**b.** 6.1 − 2.36
	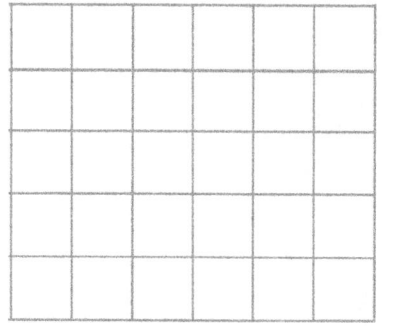

Answer Key

Addition, Subtraction, and Algebraic Thinking Review, p. 7

1.

a. $81 - 72 = 9$ $665 - 99 = 566$	b. $45 + 65 = 110$ $196 + 99 = 295$	c. $160 + 280 = 440$ $54 - 28 = 26$

2. $x + 38 = 230$
$x = 192$

← total **230** →

192	38

3. $x + 587 = 1,394$
$x = 1,394 - 587 = 807$

4. a. 30, 70   b. 100, 29   c. 82, 76

5. $(\$13 - \$2) \times 3 = \$33$.

6. $(10 \times 4) + (20 \times 2) = 80$ feet altogether.

7. $\$25 + \$14 + \$3 = \$42$

8. $\$15.20 + \$34.60 + \$70.20 = \$120$

9. $\$48.90 + (\$48.90 + \$25) = \$122.80$

Addition, Subtraction, and Algebraic Thinking Test, p. 9

1. $x = 2,611$

2. a. 260   b. 20   c. 70

3. The expression $\$20 - 7 \times \$2$ matches the problem. The answer is $6.

4. Estimations may vary because estimation is not an exact "science".
   For example: $\$30 + 2 \times \$14 = \$58$ (rounding the first number up to $30 and the other down to the nearest dollar)
   or $\$29 + 2 \times \$14 = \$57$ (rounding everything to the nearest dollar).

5. $10 \text{ m} + 8 \text{ m} + 15 \text{ m} = 33 \text{ m}$

6. $\$67 + \$48 = x$; $x = \$115$.

← original price x →

$67	$48

Large Numbers and Place Value Review, p. 10

1. a. 13,094   b. 306,050   c. 1,000,000

2. a. 785,300   b. 70,008

3. a. three thousand   b. thirty   c. 300 thousand   d. 30 thousand

Large Numbers and Place Value Review, cont.

4.

n	78,974	5,367	2,558	407,409	299,603
rounded to nearest 1,000	79,000	5,000	3,000	407,000	300,000
rounded to nearest 10,000	80,000	10,000	0	410,000	300,000

5. Estimate: 5,100 − 2,800 − 700 = 1,600. Exact: 1,556.

6. a. 500   b. 700,000   c. 600,000

7. a. 5,406 < 5,604
   b. 49,530 < 49,553
   c. 605,748 > 60,584

8. 95,695 < 145,900 < 495,644 < 496,455 < 590,554 < 5,905,544

9. a. 392,054
   b. 444,869

10. 500 × $100 = $50,000

11. In ten months, Mark earns 10 × $2,560 = <u>$25,600</u>.
    In two months, he earns $2,560 + $2,560 = <u>$5,120</u>.
    In 12 months, he earns $25,600 + $5,120 = <u>$30,720</u>.

Large Numbers and Place Value Test, p. 12

1. a. 400,040   b. 64,500   c. 200,067

2. a. eighty thousand or 80,000   b. eighty or 80

3. a. 516,800   b. 293,000   c. 200,000

4. 207,698

5. 3,294   39,244   39,294   93,294   399,295

6. The king of Nootyland has more coins; he has 5,218 coins more. The king of Sookiland has 3 × 24,000 + 1,382
   = 73,382 coins. The king of Nootyland has 78,600 coins; the difference is 78,600 − 73,382 = 5,218.

7. A thousand students.

Mixed Review 1, p. 13

1. Number sentence: $23.50 + $19.90 + $6.60 = x
   x = $50
   He paid with a $50-dollar bill.

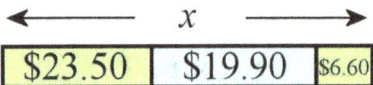

2. a. $56.25   b. $283.01

3. The total cost is ($15 − $3) × 4.

4. 5,209 < 25,539 < 25,925 < 525,009

5. a. x = 47
   b. x = 1,266
   c. x = 633

6. a. $158 − $38 = $120. Grandma gave him <u>$120</u>.
   b. 3 × $4 + $28 = $40. Jill's birthday money was <u>$40</u>.
   c. $60 − 2 × $11 = $38. He had <u>$38 left</u>.
   d. 3 × $0.60 + $0.80 = $2.60. His total cost was <u>$2.60</u>. He received $10 − $2.60 = <u>$7.40</u> in change.

Mixed Review 1, cont.

7. a. 60,000 + 8,000 + 50 + 6   b. 800,000 + 10,000 + 5,000 + 200 + 20 + 4

8. a. $1.00   b. $8.00   c. $35.00   d. $166.00
   e. $95.00   f. $99.00   g. $100.00   h. $101.00

9. a. $20 + $15 + $25 = $60.
   b. 4,000 × $1 + 1,000 × $1 = $5,000

Mixed Review 2, p. 15

1. a. 138; 74; 103   b. 58; 92; 144   c. 127; 70; 144

2. a. 1,800; 1,000   b. 600; 510   c. 9; 1   d. 500; 140

3. a. Continue this pattern: subtract _80_ each time.

700	620	540	460	380	300	220	140

b.

0	99	198	297	396	495	594	693

4. Addition: 450 + 128 + x = 1,000; Solution x = 1,000 − 128 − 450 = 422

5. a. 4,445   b. 13,378   c. 716,051

6. a. 8,030 < 18,399 < 818,939 < 819,090

   b. 5,220 < 52,200 < 250,500 < 520,500

7. a. 284 thousand 1   b. 50 thousand 50

2	8	4,	0	0	1		5	0,	0	5	0

8. a. 2,000   b. 10   c. 500,000   d. 40,000

9. a. 7,500   b. 2,700   c. 4,000   d. 400   e. 56,300   f. 293,600

10. $176 + $25 + $30 = $231.

Multi-Digit Multiplication Review, p. 17

1. a. 1,200; 180
   b. 4,200; 3,300
   c. 81,000; 40,000

2. a. 80; 7   b. 4; 400   c. 300; 80

3. a. 40   b. 2   c. 40

4. In about 8 weeks. 8 × $500 = $4,000.

5. a. Estimation: 7 × 50 = 350; Exact: 336
   b. Estimation: 6 × 800 = 4,800; Exact: 4,878
   c. Estimation: 20 × 20 = 400; Exact: 378
   d. Estimation: 4 × 6,000 = 24,000; Exact: 23,612

6.
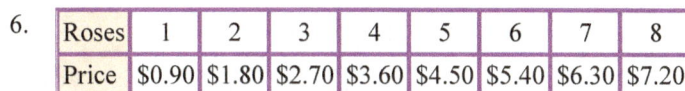

Roses	1	2	3	4	5	6	7	8
Price	$0.90	$1.80	$2.70	$3.60	$4.50	$5.40	$6.30	$7.20

7. 2 × 98 = 196; 8 × 17 = 136; 196 − 136 = 60

8. a. 2,000   b. 0   c. 80   d. 20,000

Multi-Digit Multiplication Review, continued

9.

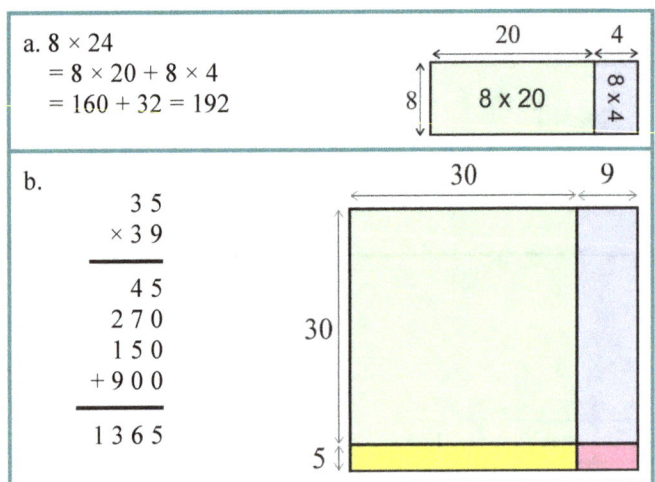

a. 8 × 24
= 8 × 20 + 8 × 4
= 160 + 32 = 192

b.
```
    3 5
  × 3 9
  ‾‾‾‾‾
    4 5
  2 7 0
  1 5 0
+ 9 0 0
‾‾‾‾‾‾‾
1 3 6 5
```

10. a. He has 50 × 20 = 1,000 shirts. The cost is 1,000 × $2 = $2,000.
   Or, write a single number sentence 50 × 20 × $2 = $2,000.

   b. 8 × $2.35 = $18.80; $20 − $18.80 = $1.20. Or, $20 − 8 × $2.35 = $1.20.

   c. 5 × $1.50 + $12.50 = $20

   d. $45 − $8 = $37. 5 × $37 = $185. Or, 5 × ($45 − $8) = $185

11. a. Two miles.

Minutes	Miles
5	1
10	2
15	3

   b. They would weigh 600 g.

Cans	Weight
1	60 g
7	420 g
10	600 g

Multi-Digit Multiplication Test, p. 20

1. a. 40 + 32 = 72   b. 140 + 42 = 182   c. 2,100 + 27 = 2,127

2. About 7 × $20 = $140.

3. a. 6,300   b. 6,000   c. 160,000

4. a. 200   b. 200   c. 8

5. a. 18,000   b. 48,000   c. 1,293   d. 2,080

6. a. 1,170   b. 5,848   c. 1,045   d. 15,924

7. 1,360

8. a. One meal costs $3, so seven meals cost $21.
   b. $30 − 7 × $2.55 = $30 − $17.85 = $12.15
   c. She took 3 × $12.55 + $8.90 + $13.45 = $60.
   d. It would have cost $30 ($150 ÷ 5 = $30).

Mixed Review 3, p. 22

1. a. 1,400   b. 4,000   c. 5,200   d. 340   e. 200   f. 9,000

2.

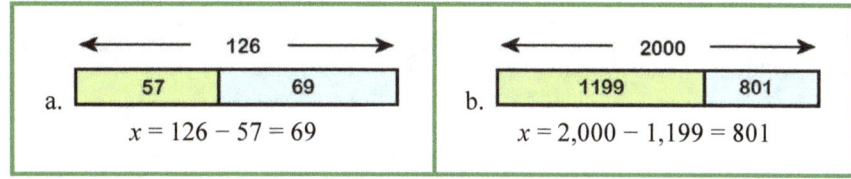

a.	126		b.	2000	
	57	69		1199	801
	$x = 126 - 57 = 69$			$x = 2,000 - 1,199 = 801$	

3. a. 440,000   b. 220,000   c. 617,100
   d. 200,000   e. 300,000   f. 81,000

4. a. 7,600; 4,000
   b. 6,190; 26,700
   c. 98,000; 430,000

5. a. 156; 80; 84
   b. 70; 50; 64
   c. 21; 9,990; 1,000

6. a. 119; 980
   b. 0; 700

7. a. <   b. =   c. <

8. Each time, Mason forgets to add the regrouped tens digit. Correct answers: 336, 384, 717.

9. a. 3 × $345 + $345 = $1,380 total
   b. $145,600 + $12,390 = $157,990. $157,990 × 3 = $473,970 total costs for June, July, and August.
      No, the total cost is not more than half a million dollars.

Mixed Review 4, p. 24

1. a. 9; 59; 590; 190   b. 6; 66; 160; 600   c. 5; 75; 500; 550

2. $458 + $366 + $427 + $503 + $413 = $2,167

3. a.

Kilometers	80	400	560	720	800	960	1,200	1,600
Hours	1	5	7	9	10	12	15	20

b.

Dollars	$9	$18	$27	$36	$45	$72	$90	$135
Yards	1	2	3	4	5	8	10	15

4. a. 1,554 < 5,000 < 5,005 < 5,500 < 5,604

   b. 3,800 < 37,700 < 38,707 < 73,737 < 307,988

5. a. 983,177   b. 555,330

6. a. $3.05 ≈ $3.00   b. $8.32 ≈ $8.00   c. $25.97 ≈ $26.00

7. a. Approximately $130 + $75 + $90 + $140 + $70 = $505.
   b. They earned about $65 less on their worst day than on their best day. (Worst day: about $75, best day about $140.)

8. a. He read 12 × 96 = 1,152 pages.
   b. Half of the magazines is 6 magazines. Jesse read six magazines in 6 × 2 1/2 hours = 15 hours.

Time and Measuring Review, p. 26

1. a. 6 hours and 52 minutes
   b. 10 hours and 40 minutes

2. The plane lands at 6:10 p.m.

3. a. A cold winter day.   b. A nice temperature for indoors.

4. Check the student's work.

   a. 2 3/8 in

   b. 36 mm

5. a. 150 mm; 68 mm   b. 12 ft; 68 in   c. 425 cm; 8,000 m

6. 16 ft 10 in

7. 5 km 600 m total per week

8. a. 16 kg or 34 lb   b. 2 kg   c. 2 oz

9. a. 112 oz, 91 oz
   b. 6,200 lb, 7,500 g
   c. 2,500 g, 3 kg 456 g

10. He weighed 20 kg 850 g before.

11. The cat food will last five days with 2 ounces left over.

12. a. 3 gal   b. 3 cups   c. 1/2 gal

13. a. 2,300 ml, 6 L 550 ml
    b. 6 pt, 12 cups
    c. 16 qt, 16 fl oz

14. It provided 48 cups of punch. (Three gallons is 12 quarts, and that is 48 cups.)

Puzzle corner: She paid $64.00. A quart has 4×8 oz = 32 oz. The total cost is then 32 oz × $2/oz = $64.

Time and Measuring Test, p. 28

1. 3:50 p.m.

2. a. 2 3/8 in. or 6 cm 0 mm   b. 3 7/8 in. or 9 cm 8 mm

3.

a.	b.	c.
4 lb 2 oz = 66 oz	2 L 80 ml = 2,080 ml	7 m 5 cm = 705 cm
76 cm = 760 mm	3 qt = 12 cups	4 kg 500 g = 4,500 g
5 ft 5 in = 65 in	200 yd = 600 ft	3 T = 6,000 lb

4. 10 cm 4 mm

5. a. two bottles   b. five bottles.

6. a. $7.60   b. $10.50

7. eight jars

Mixed Review 5, p. 29

1. $24 \times 36 = 20 \times 30 + 20 \times 6 + 4 \times 30 + 4 \times 6 = 600 + 120 + 120 + 24 = 864$ square units.

2. a. Estimation: $60 \times 30 = 1{,}800$; Answer: 1,798.
   b. Estimation: $400 \times 8 = 3{,}200$; Answer: 3,320.
   c. Estimation: $57 \times 100 = 5{,}700$; Answer: 5,643.
   d. Estimation: $7 \times 700 = 4{,}900$; Answer: 4,669.

3. a. $3,120   b. 240 km

4. a. 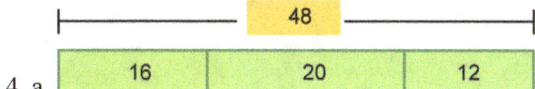 $16 + 20 + 12 = x$; $x = 48$

   b. 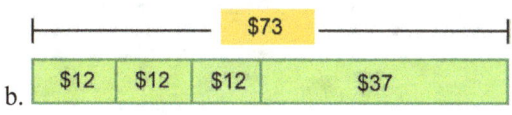 $3 \times \$12 + x = \$73$; $x = \$37$

5. a. They sold the most strawberries during week 26. About $4,500.
   b. They sold the least strawberries during week 23. The sales were $1,500.
   c. About $12,000.

Mixed Review 6, p. 31

1. a. 975   b. 20,990   c. 1,968   d. 4,088

2. $256 + x = 609$; $x = 353$.

3. a. 27   b. 57   c. 22   d. 57   e. 21   f. 1

4. a. $555 \approx 600$ b. $8{,}889 \approx 8{,}900$ c. $351{,}931 \approx 351{,}900$
   d. $64 \approx 100$ e. $244{,}295 \approx 244{,}300$ f. $38{,}009 \approx 38{,}000$

5. a. 305,200   b. 40,033

6.

a. $723{,}050 > 699{,}099$	b. $322{,}320 < 322{,}322$
c. $692{,}159 < 692{,}196$	d. $140{,}000 > 14{,}100$
e. $113{,}999 < 115{,}399$	f. $836{,}496 > 88{,}482$

7. a. 1,100; 190; 120,000
   b. 180; 800,000; 8,800
   c. 92,000; 64,000; 8,800

8.

a. $6 \times 30¢$ $= 180¢ = \$1.80$	b. $5 \times 84¢$ $= 400¢ + 20¢ = \$4.20$
c. $6 \times \$1.70$ $= \$6.00 + \$4.20 = \$10.20$	d. $3 \times \$2.80$ $= \$6.00 + \$2.40 = \$8.40$

9. a. about $20 \times 40 = 800$ plants
   b. The cost was $7 \times \$8.20 = \$56 + \$1.40 = \$57.40$. His change was $\$100 - \$57.40 = \$42.60$.

1.

a.	b.	c.
$20 \div 10 + 15 = 17$ $20 \times 10 + 15 = 215$	$(200 + 100) \div 5 = 60$ $200 + 100 \div 5 = 220$	$10 \times 12 + 40 \div 10 = 124$ $10 \times (12 + 40) \div 10 = 52$

2.

a. $3,100 \div 100 = 31$ $450 \div 10 = 45$	b. $240 \div 20 = 12$ $800 \div 40 = 20$	c. $4,200 \div 600 = 7$ $3,200 \div 80 = 40$

3.

a.	b.	c.
$45 \div 6 = 7$ R3 $46 \div 6 = 7$ R4	$12 \div 7 = 1$ R5 $27 \div 8 = 3$ R3	$31 \div 4 = 7$ R3 $56 \div 9 = 6$ R2

4. a. 236   b. 188

5. a. 78 R2   b. 474 R1

6. $288 \div 4 = 72$. Timmy has 72 seashells.

7. a. $70 \div 12 = 5$ R10. Mark had five full boxes of candles.
   b. One box had ten candles.

8. $\$38.88 \div 4 = \$9.72$. One yard cost $9.72.

9. $(92 + 85 + 89 + 75 + 89) \div 5 = 86$. John's average score was 86.

10.

Number	13	40	57	135	354	2,380
Divisible by 3			X	X	X	
Divisible by 5		X		X		X
Divisible by 10		X				X

11.

a. Is 7 a factor of 64?
   No, because it does not divide evenly into 64.
   OR No, because $64 \div 7 = 9$ R1; there is a remainder.
	b. Is 98 a multiple of 2?
	   Yes, because it is an even number.
	   OR Yes, because $2 \times 49 = 98$.

c. Is 76 divisible by 8?
   No, because $76 \div 8 = 9$ R4; the division is not even.
	d. Is 30 a factor of 30?
	   Yes, because $1 \times 30 = 30$.

12.

a. 87 is composite. $87 = 3 \times 29$	b. 89 is prime.	c. 91 is composite. $91 = 7 \times 13$

13.

a. factors: 1, 2, 3, 4, 6, 8, 12, 24	b. factors: 1, 3, 9, 27
c. factors: 1, 2, 3, 6, 11, 22, 33, 66	d. factors: 1, 3, 5, 15, 25, 75

Puzzle corner:
5, 11, 17, 23, 29, 35, 41, 47, 53, 59, 65, 71, 77, 83, 89, 95

Division Test, p. 36

1. a. 3 R1; 2 R3
   b. 4 R5; 5 R5
   c. 3 R4; 7 R4

2. One meter costs $6, so five meters would cost 5 × $6 = $30.

3. $210. One-fifth of $350 is $70; three-fifths of that is 3 × $70 = $210.

4. 250 bricks. One-third of his 1,200 bricks is 400 bricks, and two-thirds of them is double that, or 800 bricks. So, he has 400 bricks left. After selling 150 bricks, he has 250 bricks left.

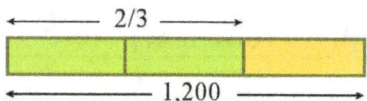

5. a. 113. Check: 5 × 113 = 565
   b. 458   Check: 8 × 458 = 3,664

6. a. 1, 2, 4, 7, 14, 28   b. 1, 13
   c. 1, 2, 4, 8, 16, 32   d. 1, 2, 4, 19, 38, 76

7. 125 ÷ 7 = 17 R6. Each child got 17 pencils, and 6 pencils were left over.

8. $47. Add the prices and divide by four: average = ($39 + $45 + $63 + $41) ÷ 4 = $47.

9. Yes, it is, because 924 ÷ 7 = 132 and there is no remainder; the division is even.

10. (30 − 10) × 20 = 400

Mixed Review 7, p. 38

1. a. 4,284   b. 49,068

2. a. 84; 80   b. 20; 54   c. 1,090; 90

3.

Estimate:	Exact: 6,859
1,568 + 4,839 + 452 ↓ ↓ ↓ ≈ 1,600 + 4,800 + 500 = 6,900	

4. a. 1,998; 3,960; 3,991
   b. 6,990; 9,970; 991
   c. 1,900; 6,700; 9,400

5. a. 34,268
   b. 800,046
   c. 406,780

6.

a. 3 ft = 36 in
   9 ft = 108 in
	b. 2 ft 5 in = 29 in
	   7 ft 8 in = 92 in
		c. 9 ft 2 in =110 in
		   10 ft 11 in = 131 in

7. a. 5,400 = 90 × 60   b. 16 × 20 = 8 × 40

   c. 7 × 49 + 49 = 8 × 49   d. 24,000 = 300 × 80

   e. 7 × 13 = 5 × 13 + 26   f. 1,500 − 500 = 5 × 200

Mixed Review 7, cont.

8. a. Estimate: 8 weeks (8 × $50 = $400). Exact: 9 weeks, because 9 × $45 = $405. He will have $6 left over.
   b. She needs 230 cm of string, 69 sheets of paper, and 46 rolls of tissue paper.

   c. James had 25 marbles.

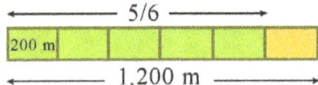

Mixed Review 8, p. 40

1. a. 1,000 meters are not accessible by boat.

   (Each little "block" in the diagram is 200 m.)

   b. There are 22 girls. There are 33 boys
      and girls.

2. a. The baker spent about $90 more for flour in May than in March.
      In May, he spent about $550 and in March, about $460.
   b. Estimates may vary. He spent about $460 + $620 + $550 = $1,630.

3. a. One triangle weighs 3 units. Solution: First take off two triangles from both sides.
      That leaves: 11 = 8 + triangle. So, one triangle has to equal 3.

   b. One square weighs 2 units. Solution: Take off one square from both sides.
      That leaves: 3 squares + 9 = 15. Now, take off "9" from both sides.
      That leaves 3 squares = 6. So, one square has to weigh 2.

4. a. 1 kg 300 g = 1,300 g   b. 3 lb = 48 oz   c. 7,500 g = 7 kg 500 g
      4 kg 20 g = 4,020 g   7 lb = 112 oz   4 lb 8 oz = 72 oz

5.

Minutes	1	5	6	7	10
Seconds	60	300	360	420	600

Days	1	3	6	10
Hours	24	72	144	240

6. First do 5 + 39 , which equals 44 . Then, divide that answer by 4 .

   This leaves 11 . Then, do 2 × 2 = 4 .
   Lastly, subtract that from 11 . The answer is 7 .

7. a. 4 h 54 m   b. 7 h 24 m   c. 8 h 24 m

8.

a.	b.	c.	d.
2 pt = 4 C	1 qt = 4 C	6 L = 6,000 ml	2 L 560 ml = 2,560 ml
2 C = 16 oz	2 gal = 8 qt	1/4 L = 250 ml	1,300 ml = 1 L 300 ml

9. a. January, February, March, April, September, October, November, and December
      (the temperature is below freezing, or below 32°).
   b. November, December, and January
   c. 37°

1. Side 1: 17 1/2 ft; Side 2: 10 ft; Side 3: 17 1/2 ft

2. The area is 320 m².
   First, divide the shape into two rectangles. You
   need to use subtraction to find some missing lengths
   of sides.

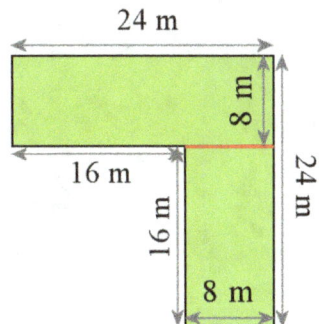

   The upper rectangle is 24 m × 8 m so its area is 192 m².
   The lower rectangle is 8 m × 16 m so its area is 128 m².
   In total, the area is <u>320 m²</u>.

3. a. 100°   b. 33°

4. a. right   b. acute   c. obtuse   d. acute   e. acute
   f. right   g. right   h. obtuse   i. right   j. obtuse
   k. obtuse

5. Check the student's answer. Here is one such angle:

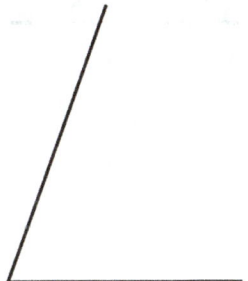

6. It is 84°. The student can figure it out in any manner.
   If we use an equation, we write:
   64° + x + 32° = 180°;   x = 84°.

7.

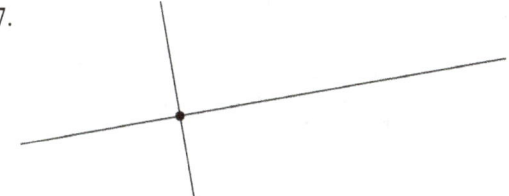

8. Task b. is possible to do. Answers will vary as the other
   two angles can vary. For example:

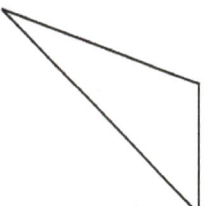

9. Check the student's drawing.

10. They are right triangles.

11. $\overline{AB} \parallel m$, $\overline{BC} \parallel n$,
    $\overline{AC} \perp m$, $\overline{AB} \perp \overline{AC}$

12. Answers will vary. Check the student's drawing.

13. a.

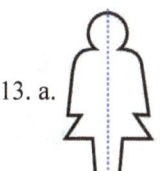

   b. not symmetrical

   c.

   d. not symmetrical

   e.

Geometry Test, p. 46

1. Check the student's answers. Here is a 75° angle:

2. a. 140°

3. It is 148 degrees.
   $78° + 134° + x = 360°$
   $x = 360° − 134° − 78°$
   $= 148°$

4. a. A parallelogram or a rhombus (either is correct).

   b.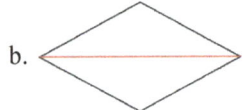

   c. 17 cm 7 mm

5. Answers vary. Check the student's answers. For example:

6. Answers vary. Check the student's answers. The two other angles should have an angle sum of 90°.

7. a. A right triangle.   b. An acute triangle.

8. Subtract the area of the outer rectangle and the area of the white rectangle:
   20 ft × 10 ft − 10 ft × 3 ft = 200 ft² − 30 ft² = 170 ft².

Mixed Review 9, p. 48

1. a. 3,000 g;   7,400 g
   b. 5,000 ml;   2,060 ml
   c. 9,000 m;   4,250 m

2. There are 1 1/2 liters left now. Six glasses got filled.

3. a. 48 in; 74 in
   b. 24 fl oz; 8 qt
   c. 64 oz; 121 oz

4. The 1-pint bottle is more. It is 4 ounces more.

5. a. 8 in   b. 24 in or 2 ft   c. 8 ft

6. a. 960   b. 508   c. 1,670   d. 1,099

7. a.

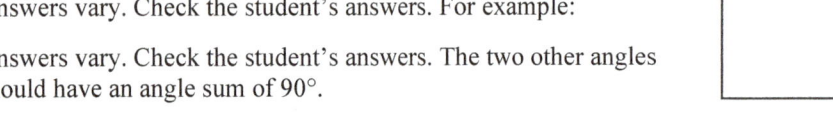

Weight (ounces)	Frequency
83..88	3
89..94	6
95..100	6
101..106	3
107..112	1
113..118	1

b. The average is <u>95 1/2 ounces</u>. You can see it in the bar graph because the number 95 1/2 is near the middle and near the peak of the graph. You could also see it from the data itself, noting that lots of the weights are 90-something.

Mixed Review 9, cont.

8. a. The silverware set costs $4 × $13 = $52. The two items together cost $13 + $52 = <u>$65</u>.
   b. He still has $200 − 8 × $18 = <u>$56</u>.
   c. From 22:15 p.m. till 7:00 a.m. is 8 hours 45 minutes. But he did not sleep from 3:30 till 5:10, which is 1 hour
      40 minutes. So, we subtract those two mounts and get that he slept 8 h 45 min − 1 h 40 min = <u>7 h 5 min</u>.

Mixed Review 10, p. 50

1. a. $18 + x = $33; $x = $15.$ The unknown x is how much Dana earned.
   b. $100 − $86 = x; $x = $14.$ The unknown x is Dad's change.
   c. $120 − 39 = x; x = 81.$ The unknown x is the number of eggs that broke.
   d. $13 + 43 = x; x = 56.$ The unknown x is the number of dogs the shelter had initially.

2. a. 590. Check: 590 × 3 = 1,770
   b. 878. Check: 878 × 9 = 7,902.

3. a. 5 R3; 2 R9   b. 5 R3; 3 R8   c. 9 R2; 11 R1

4.

a. 7 × 78	b. 13 × 67	c. 311 × 8
≈ 7 × 80 = 560	≈ 13 × 70 = 910 OR	≈ 300 × 8 = 2,400 OR
	≈ 10 × 70 = 700	≈ 310 × 8 = 2,480

5. a. 60,000 + 70 = 60,070   b. 123,000 + 4,000 + 4 = 127,004

   c. 3 + 90,000 + 40 = 90,043   d. 7 + 20 + 632,000 = 632,027

6.

a. 7 m = 700 cm	b. 2 m 6 cm =206 cm	c. 4 km 100 m = 4,100 m
69 mm = 6 cm 9 mm	6 km = 6,000 m	169 cm = 1 m 69 cm

7.

a. 3 lb 8 oz = 56 oz	b. 32 oz = 2 lb	c. 7 lb 2 oz = 114 oz
4 kg 11 g = 4,011 g	4,900 g = 4 kg 900 g	36 kg 140 g = 36,140 g

8. a. $205
   b. 2 km 600 m
   c. She spent $7.12. Her change was $2.88

Fractions Review, p. 52

1. a. 1   b. 5 1/8   c. 7
   d. 2/10   e. 1 2/4   f. 6 7/12

2. a. 7/10   b. 3/5   c. 4/5   d. 5/8

3. a. 33/100   b. 53/100   c. 1 17/100

4. 1 3/4 liters

5. Answers vary. For example: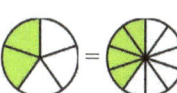

6.

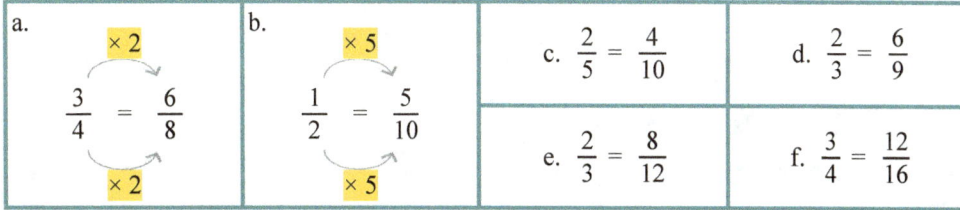

Fractions Review, cont.

7. a. > b. < c. = d. >
   e. > f. > g. > h. <

8. a. 9/10 b. 1 1/5 c. 1 4/10
   d. 99/100 e. 2 4/8 f. 2 9/12

9.

Mexican Coffee (4x)
6 cups strong gourmet coffee
3 tsp cinnamon
16 tsp chocolate syrup
1 tsp nutmeg
2 cup heavy cream
4 tbsp sugar

10.

a. 40 80	b. 8 cm 40 cm	c. 400 kg $40

11. Since he has 1/4 of his birthday money left, he has $\underline{\$5}$ left.

12. One-eighth of 240 pages is 30 pages. She has 5/8 of the book left to read, which is $5 \times 30 = \underline{150 \text{ pages}}$.

Fractions Test, p. 54

1. a. 1 b. 2 1/3 c. 5 4/5
   d. 2/12 e. 2 3/5 f. 5 4/6

2. a. $\dfrac{3}{8}, \dfrac{1}{2}, \dfrac{3}{4}$ b. $\dfrac{5}{7}, \dfrac{5}{5}, \dfrac{7}{5}$ c. $\dfrac{5}{9}, \dfrac{5}{6}, \dfrac{5}{2}$

3.

a. Split all pieces into four new ones.	b. Split all pieces into three new ones.
$\dfrac{1}{2} = \dfrac{4}{8}$ 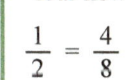	$\dfrac{2}{3} = \dfrac{6}{9}$

4.

a. $\dfrac{1}{5} = \dfrac{2}{10}$	b. $\dfrac{3}{4} = \dfrac{9}{12}$	c. $\dfrac{4}{5} = \dfrac{20}{25}$	d. $\dfrac{1}{6} = \dfrac{4}{24}$

5. a. 12/10 = 1 2/10   b. 15/5 = 3   c. 12/8 = 1 4/8 (which is equal to 1 1/2)

6. a. Walter and Eric ate equal amounts. Walter ate 1/4 of it, which is equal to 3/12, or 3 pieces.
   b. Eric ate 3/12 and John ate 1/12. So, Eric ate 2/12 of the pizza more than John

Mixed Review 11, p. 55

1.

a. 57 ÷ 5 = 11 R2 11 × 5 + 2 = 57	b. 34 ÷ 7 = 4 R6 4 × 7 + 6 = 34	c. 33 ÷ 9 = 3 R 6 3 × 9 + 6 = 33

2.

a. A = 7 in × 3 in = 21 in²	b. A = 25 km × 20 km = 500 km²	c. A = 2 ft × 9 1/2 ft = 19 ft²

Mixed Review 11, cont.

3. a. acute   b. right   c. obtuse   d. right   e. obtuse   f. acute   g. right   h right

4. a. Answers vary. Check the student's answers. For example:
   b. Answers vary. The measurements in this picture are not to scale.
   c. Answers vary. The angles of the parallelogram above are 58°, 122°, 58°, and 122°.

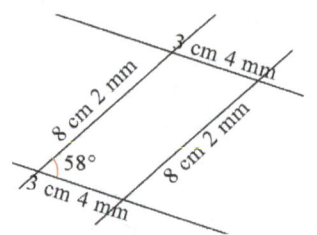

5. a. 268   b. 277

6. a. 4 × 44 lb = 176 lb and 5 × 32 lb = 160 lb.
      Four boxes 44 lb each weigh more. They weigh 16 lb more than five boxes 32 lb each.
   b. 86 ÷ 25 = 3 R11. The teacher got to keep 11 balloons.
   c. $9.73 ÷ 7 = $1.39 and 5 × $1.39 = $6.95. Five liters would cost $6.95.

7. $20 − ($7 + $5) = $8.

Mixed Review 12, p. 57

1.

a. 22,934 + 5,312 + 424,787 　Estimation: 23,000 + 5,000 + 420,000 = 448,000 　Calculation: 453,033	b. 519,313 − 47,616 　Estimation: 520,000 − 50,000 = 470,000 　Calculation: 471,697

2. 45 × 22 + 27 = 1,017 students

3. Right angles are exactly 90°.
   Right triangles have exactly one right angle.

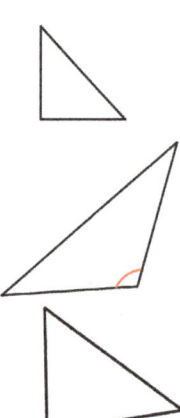

   Obtuse angles are more than 90°, but less than 180°.
   Obtuse triangles have exactly one obtuse angle.

   Acute angles are less than 90°.
   Acute triangles have three acute angles.

4. Answers vary. To find possible side lengths, remember that the two side lengths add up to 14 in.

One side	Other side	Perimeter	Area
2 in.	12 in.	28 in.	24 sq. in.
3 in.	11 in.	28 in.	33 sq. in.
4 in.	10 in.	28 in.	40 sq. in.
5 in.	9 in.	28 in.	45 sq. in.

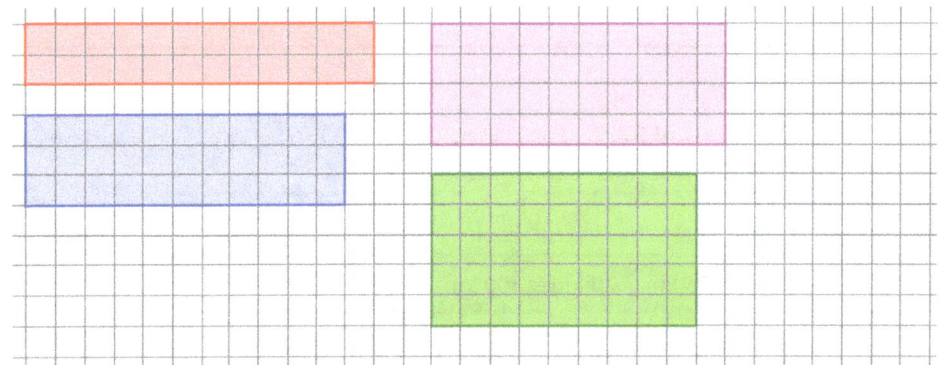

5.

a. 1:40 p.m.	b. 9:20 p.m.	c. 2:15 p.m.	d. 10:04 a.m.
_13 : 40_	_21 : 20_	_14 : 15_	_10 : 04_

6.

number	divisible			number	divisible			number	divisible		
	by 2	by 5	by 10		by 2	by 5	by 10		by 2	by 5	by 10
478	x			1,492	x			904	x		
540	x	x	x	3,093				905		x	
255		x		94	x			906	x		

7. No, because the division leaves a remainder: 549 ÷ 7 = 78 R3

8. a. 1, 2, 3, 4, 6, 8, 12, 24   b. 1, 2, 3, 6, 11, 22, 33, 66 c. 1, 2, 3, 4, 6, 8, 12, 16, 24, 32, 48, 96 d. 1, 3, 5, 15, 25, 75

9.

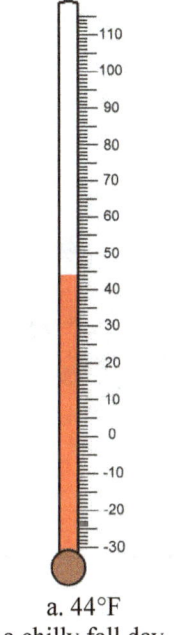 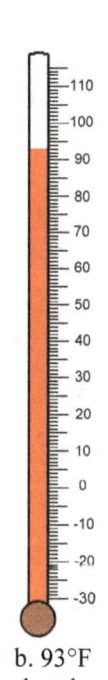 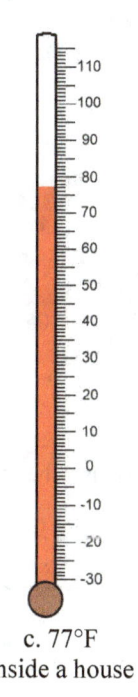 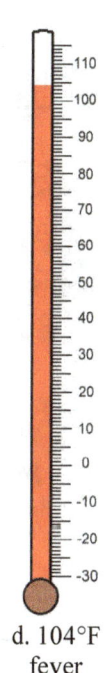 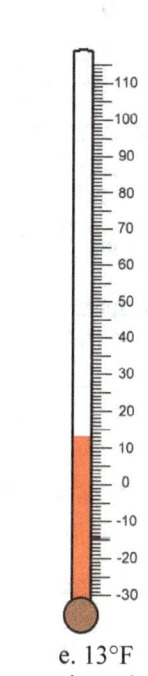

a. 44°F	b. 93°F	c. 77°F	d. 104°F	e. 13°F
a chilly fall day	a hot day	inside a house	fever	a winter day

Decimals Review, p. 59

1. a. 0.7   b. 0.07   c. 1.6   d. 2.41
   e. 1.01   f. 0.47   g. 8/10   h. 2 9/10
   i. 4 14/100   j. 18 8/100   k. 3/100   l. 29/100

2. a. >   b. >   c. <
   d. =   e. >   f. =
   g. <   h. <   i. >

3.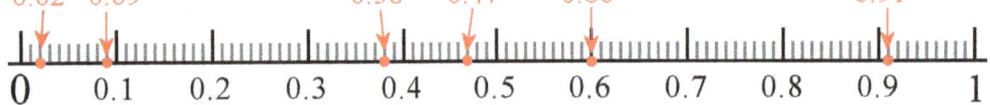

4. 0.1, 0.12, 0.2, 0.21, 1/2, 0.74, 0.8

5.

a. $0.7\underline{0} + 0.03 = 0.73$	b. $0.32 + 0.4\underline{0} = 0.72$	c. $0.7\underline{0} - 0.04 = 0.66$
$\dfrac{70}{100} + \dfrac{3}{100} = \dfrac{73}{100}$	$\dfrac{32}{100} + \dfrac{40}{100} = \dfrac{72}{100}$	$\dfrac{70}{100} - \dfrac{4}{100} = \dfrac{66}{100}$

6. a. 1   b. 0.88   c. 0.36
   d. 0.24   e. 0.83   f. 0.5

7. a. Incorrect. Should be: 0.99 + 0.1 = 1.09   OR   0.99 + 0.01 = 1.
   b. Correct.
   c. Incorrect. Should be: 0.19 + 0.19 = 0.38.
   d. Incorrect. Should be: 0.03 + 0.5 = 0.53   OR   0.03 + 0.05 = 0.08.

8. a. 9.31   b. 23.11   c. 5.84

9. 2.84

10. 900 grams; 200 grams; a tablet weighing 610 grams is heavier

Decimals Test, p. 62

1.

2. a. 0.2   b. 7.04   c. 0.74   d. 52/100   e. 3 9/10

3. a. 2.2   b. 0.95   c. 0.19
   d. 0.7   e. 0.37   f. 3.04

4. a. >   b. =   c. >   d. <   e. <

5. 2.07 < 2.17 < 2.7 < 2.77 < 7.2

6. 5.2 kg. You can add 1.3 kg + 1.3 kg + 1.3 kg + 1.3 kg = 5.2 kg

7. a. 7.36   b. 1.76

Mixed Review 13, p. 63

1. a. 5/8 + 3/8 + 2/8 = 1 2/8
   b. 1 7/12 + 7/12 = 2 2/12

2. a. 50/12 = 4 2/12
   b. 28/9 = 3 1/9
   c. 42/100

3. a. 1, 2, 19, 38
   b. 1, 2, 4, 7, 8, 14, 28, 56
   c. 1, 19 (it is prime)

4. a. ≈ 6 × 300 = 1,800 Exact: 1,752   b. ≈ 11 × 400 = 4,400   Exact 4,422
   c. ≈ 3 × 2,400 = 7,200   Exact 7092   d. ≈ 7 × 9,000 = 63,000   Exact 61,789

5. a. 90
   b. 4
   c. 30

6. a. 2:30 pm
   b. 7:15 pm
   c. 10:45 pm
   d. 7:50 am

7.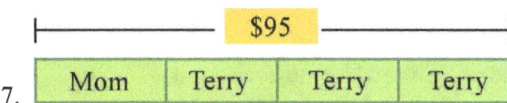

   Mom paid $23.75 and Terry paid $71.25.

8. He still has to pay $240. First find 1/5 of $600, which is $120. Jack still has to
   pay 2/5 of the price, which is 2 × $120 = $240.

9.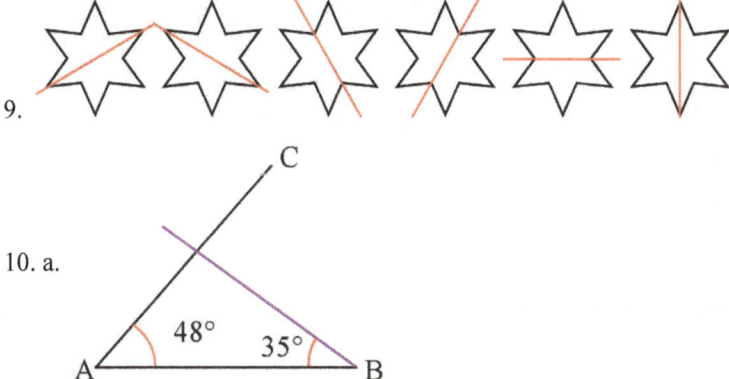

10. a.

b. 97°

Mixed Review 14, p. 65

1.

a. 4 × 36	b. 5 × 65	c. 8 × 426
120 + 24 = 144	300 + 25 = 325	3,200 + 160 + 48 = 3,408

2. 98,889

3. Answers vary since estimations can be done in various ways.

a. 8 × 69	b. 11 × 55	c. 25 × 17
≈ 8 × 70 = 560	≈ 10 × 60 = 600 OR 11 × 60 = 660 OR 10 × 55 = 550	≈ 25 × 20 = 500 OR 24 × 20 = 480 (better, rounding one down, one up)

4. a. $x \div 8 = 7$; $x = 56$
   b. $24 \div x = 8$; $x = 3$

5. yes - no - no - yes - no - yes - no

6.

a. 5 ft = 60 in 12 ft = 144 in	b. 3 ft 4 in = 40 in 6 ft 6 in = 78 in	c. 4 yd = 12 ft 9 yd = 27 ft

7. a. 5 kg 500 g
   b. 3 kg 400 g
   c. 9,900 g

8. a. Their total weight was 1 lb 1 oz. 3 oz + 3 oz + 5 oz + 2 oz + 4 oz = 17 oz = 1 lb 1 oz.

   b. 44 boxes. The division is 175 ÷ 4 = 43 R3. Keep in mind, she needs to pack the "leftover" 3 kg also into one box.

   c. She has $84 left. $98 ÷ 7 = 14; $98 − $14 = $84.

   d. Either 2, 3, 4, 6, 9, 12, or 18 children in a row. Probably 4, 6, and 9 children in a row are most practical.

   e. There are 34 foals, and 102 horses in total.

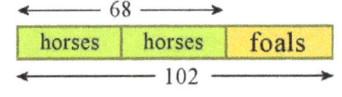

9. Check the student's work. a.

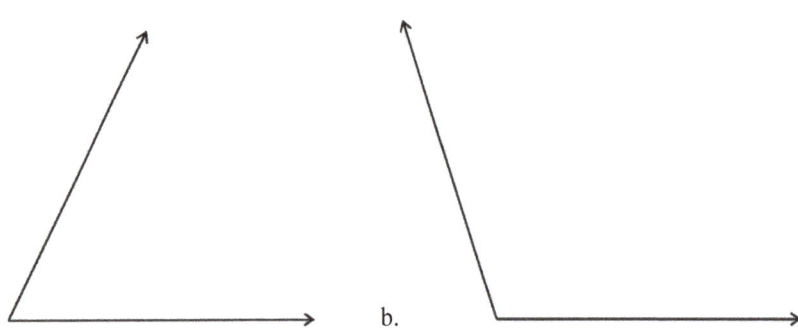

b.

10. $o \parallel AB$, $o \parallel m$, $AB \parallel m$. There are no perpendicular lines, rays, or line segments.

11. Three-fourths of it are left.

12. a. 3   b. 1 1/12   c. 13/100   d. 1 1/4   e. 26/100   f. 6 1/10

13. a. 3   b. 1 1/5   c. 3 3/4   d. 7

14.

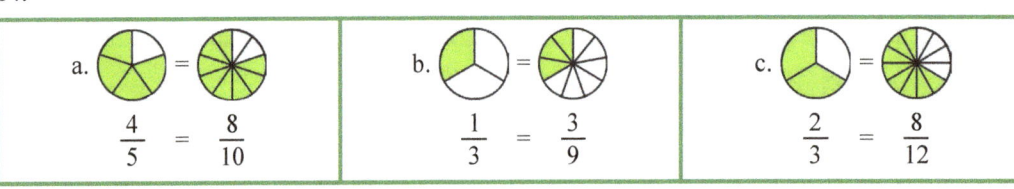

a. = $\frac{4}{5} = \frac{8}{10}$	b. = $\frac{1}{3} = \frac{3}{9}$	c. = $\frac{2}{3} = \frac{8}{12}$

15. a. <   b. <   c. <   d. >

1. 1,980. Add to check: 1,980 + 543 + 2,677 equals 5,200.

2. a. ≈ $1 + $9 + $4 + $9 = $23
   b. Her bill is $1.28 + $8.92 + $3.77 + $9.34 = $23.31. Her change is $30 − $23.31 = $6.69.

3. Estimate: 5 × $0.90 + 2 × $1.20 = $4.50 + $2.40 = $6.90

4. a. 30; 84   b. 11; 14   c. 140; 19

5. a. $35 + x = $92 ; x = $57
   b. x − 24 = 37; x = 61

6. a. 2,000 1,750 1,500 1,250 1,000   750   500   250

   b. 200, 500, 800, 1100, 1400, 1700

7. In the frequency table we list how many students got that score.

Quiz score	Frequency
1	0
2	1
3	0
4	1
5	3
6	5
7	5
8	4
9	3
10	2

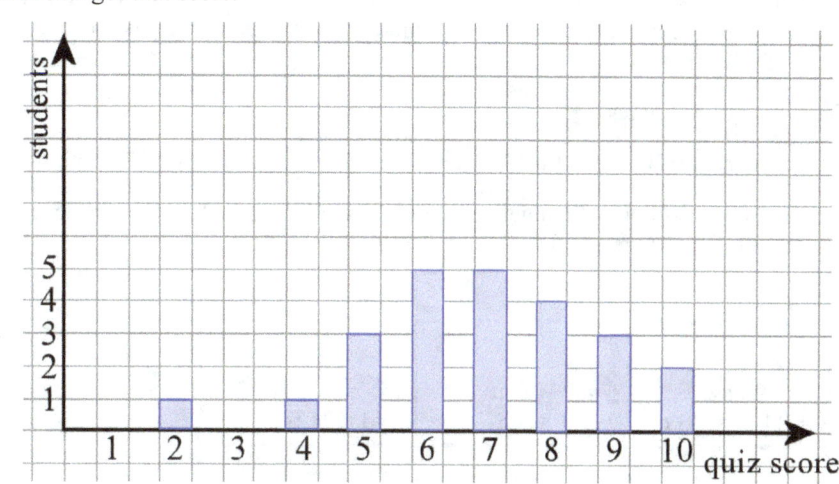

8.

Rubber boots used to cost $27.95 but now the price is $21.45. How much is the discount?
$21.45 + x = $27.95   OR x = $27.95 − $21.45
x = $6.50

←—— original price $27.95 —→

| $21.45 | x |

9. a. 1,999   b. 4,980   c. 5,700

10. a. 800,050   b. 25,407

11. a. 30,000   b. 9,000   c. 600

12. a. <   b. >   c. >

13. 27,200   217,200   227,200   227,712

14. a. 440,000   b. 90,000   c. 27,500

15. a. 430,000   b. 500,000   c. 10,000

16. a. 501,663   b. 323,688

17. a. 210   b. 4,800   c. 3,200   d. 120   e. 80   f. 70

18. a. $160
    b. $800
    c. four days, since 4 × $160 = $640

19. a. estimate 5 × 200 = 1,000. Exact: 980
    b. estimate 40 × 40 = 1,600 or 30 × 40 = 1,200. Exact: 1,330
    c. estimate 7 × 3,000 = 21,000. Exact: 22,316
    d. estimate 90 × 20 = 1,800. Exact: 1,958

20.

Area = 8 × 127
= _8_ × _100_ + _8_ × _20_ + _8_ × _7_
= 800 + 160 + 56 = 1,016

21. a. Answers may vary. For example: $400 − 26 × $14 = $400 − $364 = $36. Or, 26 × $14 = $364 and $400 − $364 = $36.
    b. 24 × 60 minutes = 1,440 minutes
    c. Answers may vary. For example: 4 × 375 cm = 1,500 cm. Or, 375 cm + 375 cm + 375 cm + 375 cm = 1,500 cm
    d. Answers may vary. For example: ($277 − $58) × 8 = $1,752. Or, $277 − $58 = $219 and 8 × $219 = $1,752.

22. Answers may vary if the test is printed with "shrink to fit" or "fit to printable area", or because of slight variability in rulers, or because of measuring inaccurately. Please check the student's answers.
    a. 5 1/4 in. or 13 cm 3 mm. 13 cm 4 mm is also acceptable.   b. 3 7/8 in. or 9 cm 8 mm. 9 cm 9 mm is also acceptable.

23. 6 hours 12 minutes

24. 1 h 45 min + 50 min + 1 h 15 min + 2 h 15 min + 55 min = 4 h 180 min, which is 7 hours.

25. She worked 7 hours 30 minutes. From 7:00 am till 3:35 pm is 8 hours 35 minutes. Subtract from that 65 minutes, or 1 hour 5 minutes, to get 7 hours 30 minutes.

26.

a.	b.	c.
6 lb = 96 oz 2 lb 11 oz = 43 oz	5 gal = 20 qt 2 qt = 8 cups	4 ft 2 in. = 50 in. 7 yd = 21 ft

27.

a.	b.	c.
2 kg = 2,000 g 11 kg 600 g = 11,600 g	5 L 200 ml = 5,200 ml 3 m = 300 cm	8 cm 2 mm = 82 mm 10 km = 10,000 m

28. In four days, he jogs 15 km 200 m.

29. 1 L 650 ml

30. 17 ft 8 in

31. a. 63. Check: 63 × 9 = 567
    b. 2,141. Check: 2141 × 4 = 8,564

32. a. 9 R2   b. 8 R1   c. 6 R3

33. a. Three photos on the last page; five pages were full.
    b. Your neighbor should be $36, because one foot of the fence costs $3.

34. a. It cost $99. First find 1/8 of $264: $264 ÷ 8 = $33. Then to find 3/8 of it, multiply 3 × $33 = $99.
    b. She needs 20 bags. 117 ÷ 6 = 19 R3. Notice she needs a bag also for the three muffins that don't fill a bag.

35.

number	divisible by 1	divisible by 2	divisible by 3	divisible by 4	divisible by 5	divisible by 6	divisible by 7	divisible by 8	divisible by 9	divisible by 10
80	x	x		x	x			x		x
75	x		x		x					
47	x									

36.

a. Is 5 a factor of 60?	b. Is 7 a divisor of 43?
Yes , because _5_ × _12_ = _60_ .	_No_ , because _43_ ÷ _7_ = 6 R1 (the division is not even).
c. Is 96 divisible by 4? Yes , because _96 ÷ 4 = 24_ (the division is even).	d. Is 34 a multiple of 7? _No_ , because 34 is not in the multiplication table of 7. OR: No, because 34 ÷ 7 = 4 R6; the division is not even. OR: No, because there is no whole number you can multiply by 7 to get 34.

37. Answers vary. For example: 2, 3, and 5. Here is a list of primes less than 100:
2 3 5 7 11 13 17 19 23 29 31 37 41 43 47 53 59 61 67 71 73 79 83 89 97

38. a. 1, 2, 4, 7, 8, 14, 28, 56
   b. 1, 2, 3, 6, 13, 26, 39, 78

39. 155°

40. Check the student's answers.

41. Answers vary. Check the student's answers. The sum of the angle measures should be 180° or very close.

42. 29° + x = 180°; x = 151°.

43. Right angles.

44. Answers vary. Check the student's answers. For example:

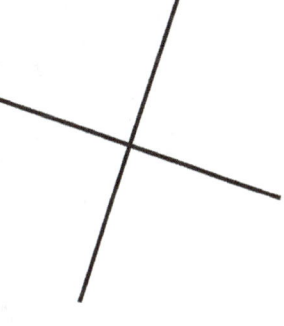

45.

46. Use subtraction. A = 28 ft × 12 ft − 6 ft × 10 ft = 336 ft² − 60 ft² = 276 ft².

47. $\dfrac{5}{8} + \dfrac{5}{8} = 1\dfrac{2}{8}$

48. There is still 2/4 or 1/2 of it left to do.

49. a. 1 2/5   b. 5/6   c. 6

50.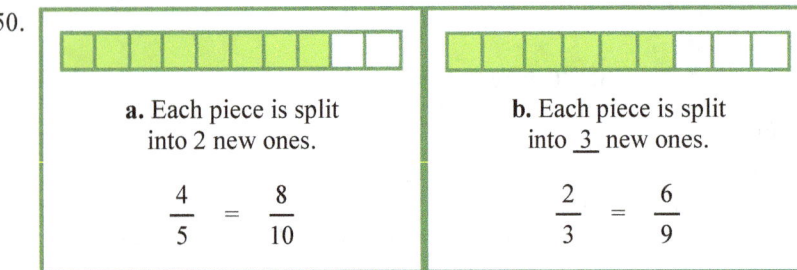

a. Each piece is split into 2 new ones.

$$\frac{4}{5} = \frac{8}{10}$$

b. Each piece is split into _3_ new ones.

$$\frac{2}{3} = \frac{6}{9}$$

51.

a. $\frac{2}{3} = \frac{10}{15}$	b. $\frac{3}{5} = \frac{9}{15}$	c. $\frac{1}{6} = \frac{2}{12}$	d. $\frac{1}{3} = \frac{3}{9}$

52. a. >   b. >   c. <   d. <

53. $\dfrac{65}{100} < \dfrac{7}{10} < \dfrac{5}{4}$

54. 2 1/4 cups

55. a. 1/8   b. 1 3/5   c. 1 2/12

56.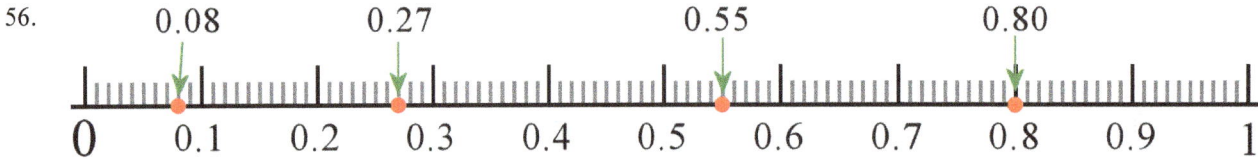

57. a. 0.3   b. 3.9   c. 0.09   d. 7.45

58. a. 6/10   b. 6 7/10   c. 21/100   d. 5 5/100

59. a. <   b. >   c. <   d. =

60. a. 13.01   b. 3.74

Math Mammoth has a variety of resources to fit your needs. All are available as economical downloads, and most also as printed copies.

- **Math Mammoth Light Blue Series**
  A complete curriculum for grades 1-7. Each grade level includes two student worktexts (A and B), which contain all the instruction and exercises all in the same book, answer keys, tests, cumulative reviews, and a worksheet maker. International (all metric), Canadian, and South African versions are also available.

  https://www.MathMammoth.com/complete-curriculum

  https://www.MathMammoth.com/international/international

  https://www.MathMammoth.com/canada/

  https://www.MathMammoth.com/south_africa/

- **Math Mammoth Skills Review Workbooks**
  These workbooks are intended to be used alongside the Light Blue series full curriculum, and they provide additional review to the topics studied in the main curriculum, in a spiral manner.
  https://www.MathMammoth.com/skills_review_workbooks/

- **Math Mammoth Blue Series**
  Blue Series books are topical worktexts for grades 1-7, containing both instruction and exercises. The topics cover all elementary mathematics from 1st through 7th grade. These books are not tied to grade levels, and are thus great for filling in gaps.
  https://www.MathMammoth.com/blue-series

- **Make It Real Learning**
  These activity workbooks concentrate on answering the question, "Where is math used in real life?" The series includes various workbooks for grades 3-12.
  https://www.MathMammoth.com/worksheets/mirl/

- **Review Workbooks**
  Workbooks for grades 1-7 that provide a comprehensive review of one grade level of math—for example, for review during school break or summer vacation.
  https://www.MathMammoth.com/review_workbooks/

Free gift!

- Receive over 350 free sample pages and worksheets from my books, plus other freebies:
  https://www.MathMammoth.com/worksheets/free

Lastly...

- Inspire4 is an inspirational website for the whole family I've been privileged to help with:
  https://www.inspire4.com

www.ingramcontent.com/pod-product-compliance
Lightning Source LLC
Chambersburg PA
CBHW081734220526
45468CB00008B/2093